DESTINATION
CAPE HORN

George Brown

Published by

The
Svengali Press
2, 2-4 Notts Ave
Bondi Beach
NSW 2026
AUSTRALIA
trilby@svengalipress.com.au
www.svengalipress.com.au

ISBN 978-0-6489227-6-6 (pbk)
ISBN 978-1923024-76-2 (ebk)
Copyright © 2023 George Brown

All rights reserved worldwide. No part of the book may be copied or changed in any format, sold, or used in any way other than what is outlined in this book, under any circumstances, without the prior written permission of the copyright-holder.

Contents

Dedication	i
Introduction	ii
Chapter 1	1
Chapter 2	4
Chapter 3	41
Chapter 4	102
Chapter 5	223
Chapter 6	233
Chapter 7	274

Dedication

IT is often said that everyone has a book in them, their life story. Well, this is my life story, or the main events anyway.

It was never going to be a conventional life for me. I was born at a time when the world was changing rapidly and there were lots of opportunities. Education was free and I did better than most would have expected. I was an avid book reader and wanted to travel. After I left school, I discovered that factories and offices were not for me. I had an independent spirit and wanted to walk my own road. I travelled to Australia not expecting that I would stay there for any length of time. In Sydney I fell in love with boats and sailing and they filled my life for the next 25 years.

I would like to mention George (Archie) Howie, we became friends in Edinburgh, and he was with me when I first bought 'Dorado', and we learned to sail together on Sydney Harbour, then the voyage to Lord Howe Is., and the outward voyage to Fiji.

Also, Alf Taylor, who owned the marina that 'Dorado called home for a number of years. Alf was a master mariner, and unsung hero of WWII. He gave me help and encouragement with boat maintenance and handling, and most importantly navigation (these were the days of sextant and nautical almanac).

Then there is Paul Ellercamp whose experience as a journalist helped me greatly in the writing of this book. I was heading in the wrong direction when Paul hauled me back into the digital world and got me going in the right direction. I owe him plenty.

In 2012 Chloe, my wife, took a photo of 'Dorado' and went to see well-known Naval Artist Ian Hansen and got him to do a painting of 'Dorado' approaching Cape Horn. A photo of that painting features on the cover of this book. Ian has kindly given me permission to use his painting in this way.

Last but not least there is Chloe my wife. We met at a time when both of our lives were at a crossroads. We got together and are still together 36 years later. She is the love of my life!

GEORGE BROWN 2023

Introduction

On December 18, 1983, a small yacht was sailing towards Cape Horn. An observer would have noticed that the weather was quite moderate for that part of the world, and the yacht was carrying less sail than might be expected for the conditions. A self-steering gear was operating and seemed to be steering the yacht straight towards Cape Horn Island. There was nobody on deck tending to the yacht; the crew must be below… Time went by, and the yacht continued on its way, helped by the strong Cape Horn current.

Disaster was looming, where were the crew? The yacht was standing on straight towards the sheer cliffs underneath the high point of the island. There were rocks and reefs lying offshore, but somehow, she missed these. No matter, her fate was sealed; she was now getting the backwash from the cliffs and being thrown about all over the place. Suddenly, a hatch on the cabin top slides open. A head appears, and shoulders, arms trying to reach out. Too late! A big wave comes from behind, picks up the yacht, and she's surfing down the wave, straight at the cliff… ssmack!!!

Hello! Yes, that was me on my yacht, 'Dorado'. I was shipwrecked on Cape Horn, and I survived to tell the story. Thirteen years of boats and sailing, including a circumnavigation of the world via the Panama Canal, says that I was an experienced sailor, and a better than average navigator. How could I allow myself to fall asleep in such a dangerous situation, and when I was about to achieve my greatest ambition?

Continual gale force westerly winds for 2-3 weeks before Cape Horn, which is standard weather for that latitude, then just as we're coming in from the deep ocean up onto the Cape Horn bank, a 55-knot gale, with confused breaking seas. The Cape Horn Bank is where the groundswell from the deep Pacific Ocean suddenly meets the shallower waters over the South American continental shelf, and in conditions like this, all hell breaks loose in the sea. You need to pay attention to deal with it. Knocked down twice, right upside down. Finally, we're up on the bank,

Introduction - Growing up in Scotland

the wind eases, the seas settle down. Check the boat, the mast, the rigging, all OK. Reset the self-steering, relax, have a cup of soup, exhausted, maybe just a couple of hours, standard practice on my voyages. But this is not the trade winds…

Chapter 1

Growing up in Scotland

My name is George Brown. I was born at Kelso Cottage Hospital in the Borders of Scotland on the 7th of May 1942. My father was away at the war, and my mother was living with her mother at 'Angraflat' near Kelso. It was built in the early 1900s when it was known as the 'Fever Hospital'. But after the discovery of penicillin, its use declined, and it was more or less abandoned when we lived there. All the buildings were substantially built in stone. There were two hospital wards, a matron's house, and some outhouses, plus the lodge where we lived; maybe about four and a bit acres surrounded by a six foot-high wall. A schoolteacher and his wife and daughter lived in the matron's house. Sandy Cow taught maths at Kelso High School, and in a few years was to play a major part in my education.

As a young kid, Angraflat was a paradise: old, abandoned buildings to explore, trees to climb, and bushes. There were red squirrels high up in the trees, and rabbits on the ground.

I was on my own a lot of the time, but it never bothered me. I've always been a bit of a loner, and I've walked my own road, as they say. Later on, in 1945, my brother was born; the war ended; my father came home; I went to school; and family and school- friends started coming round to our place. I would feel rather resentful about these strangers invading my paradise.

When the war ended in August 1945, there was a general election in Britain, won by the Labour Party. This ushered in a suite of socialist policies bringing great benefits to working class people. Free health was a great boon, including free dental for kids at school, and free education, right up to university if you were good enough. I went to school 5-10 years before the 'baby boomers' got going, and the teachers had been teaching before the war, and as consequence classes were smaller and I got a lot more personal attention than if I had been a 'boomer'.

When I was in fourth year, I saw an ad in the paper. It was

Shell oil co., looking for young people about my age to become 'Officer Cadets', to train to become officers on its oil tankers. I liked the sound of that very much. Travel the world and get paid for it. I wanted to sign up for it straight away, but my parents said, no, wait a minute; we think you can do better than that. We want to speak to the school before you make a decision. And of course, the school backed them up saying, yes, George can do better than that.

In later life, I have regretted missing that opportunity. I think it would have suited me down to the ground. I just should have asserted myself a bit more, and just gone for it. There was a lesson to be learned. I was too easy going. I knew from playing rugby that it took some incident in the game to stir me up, and then - look out! There were plenty of options. I could have gone to university and studied Geology. It was free in these days, and I would have received a scholarship. All the armed forces had officer cadet schemes, and a number of big companies had so-called sandwich courses. One of these was Ferranti, a company which had big factories in England, also in Edinburgh, where they built, and maintained aircraft radars on government contracts.

I didn't feel like I wanted to go to Uni at that stage, but I did know someone who had gone to Ferranti a couple of years earlier and was doing well. That was enough to swing the decision, and I applied to Ferranti, and was accepted to become a student apprentice. When all the new guys got together, it wasn't at Crewe Toll, but at Leith where Ferranti had taken over a bomb-damaged school to use as an induction centre for its new guys. None of us knew each other, but it didn't take long. One of the guys looked vaguely familiar, and it turned out that he came from Greenlaw and had gone to Duns High School, and of course I had played rugby against him several times. George Howie and I became good friends.

Ferranti's was not a success. It had been a mistake for the country boy to go into the city and work in a big factory, where we rarely saw the sun and worked in crowded offices with the

Chapter 1 - Growing up in Scotland

usual office politics. The winter months were especially depressing, and it wasn't long before I was starting to look for a way out. George Howie had been talking about emigrating to Australia for some time, and I was starting to think that he was serious. He had two lots of friends from around the Greenlaw area who had emigrated and were doing all right. I had met or heard of quite a few people who had gone away on the assisted passage scheme, and then came home disillusioned. Of course, the ones that came home would be disillusioned, wouldn't they, and there were many more who stayed and did well. I had been thinking of doing some travelling myself. Maybe I could go to Australia first, and if it didn't work out then I could go on elsewhere later. At least they were going to pay my airfare to get there, and that was worth something to get me started. And so, on the 4th of September 1964, I boarded a Qantas Boeing 707 jet at London Heathrow bound for Sydney.

Chapter 2

Australia. The Early Years

We arrived in Sydney on Sunday, September 6, 1964. The Immigration Department were there to meet and greet us. They were highly organised, and already had lodgings arranged for us, and taxis booked to take us there. I was sent to Bondi, to an address in Rowland Av., right next to the tramline as it wound its way down to Bondi Beach. The trams had stopped running a couple of years previously, and the track was a convenient short cut down to the beach.

The Immigration people had also organised 'hosts' to help us settle in. Ours were Larry and Jan Lennehan. They were great hosts. They went out of their way to help the new arrivals who were unfamiliar with their new surrounds. They showed us around the local pubs and clubs, shops and beaches; one day, they took us to the tennis. The Australian Open was played at that time at White City, in Rushcutters Bay, in Sydney's inner east, and most of the top players were Aussies. All the big names were there. It was a hot November day. When the tennis finished, it was decided that we would go up to Kings Cross for refreshments.

On another occasion, a Sunday morning, Larry said to us, 'We're going out to watch the 18 footers today. Would you like to come?' Yeah sure, we'd love to come,' we said. Doyle's fish restaurants are famous in Sydney. They were then, and they are today nearly 60 years later. This one was built on piles out over the water, with a pontoon on the end for ferries to tie up to. Our ferry was waiting for us to board, which we did quick smart, and off we went. It was a classic, sparkling Sydney Sunday afternoon. The sky was blue, the sun was warm, the north-east sea breeze was building, the racing was exciting, thrills and spills galore. there was something else. As I looked up and down the harbour, I could see what looked like hundreds of other sailing boats cruising around enjoying a Sunday afternoon sail. It was like my

Chapter 2 - Australia. The Early Years

eyes just opened; who are these people; how long has this been going on? These lucky people! I'd love to have my own boat to go sailing in. A seed was sown that day which didn't germinate for a few years, but from then on, I began to take a keen interest in boats and sailing. The afternoon concluded with a slap-up fish and chip meal at Doyle's. A great day was had by all.

It had been a very enjoyable few weeks but now the piper had to be paid. I was running out of money. I had to get a job.

'The Sydney Morning Herald' has thousands of jobs every week, I was told. They were right of course, and I was soon looking at lots of engineering jobs that I might apply for.

It didn't take too long, less than two weeks from memory; an ad caught my eye. 'Wanted: Work Study Engineer, Arnott's Biscuits, Family Company, Homebush'. I applied for the job and became one of a team of eight servicing a large factory with numerous production lines. I was made to feel at home and one of the team.

I finished at Arnott's in April '67. I'd been there just over three years. It wasn't that I was unhappy; I don't remember ever feeling unhappy, just restless. Something was driving me that needed to be expressed. I wanted to be able to walk my own road, to be independent. I wanted to have my own yacht and sail it round the world. I needed money, a lot more money than I had now to even get started. In my three-plus years in Australia, I'd lived and worked in Sydney, and seen very little else.

My thinking was that it was time to move on, but I definitely wanted to see Australia first, and that might take anything up to two years, so first things first.

When I left Sydney in May, I went north and kept going north. It was in Townsville that I realised that I was going to have to get a job. Mount Isa mines had an employment office in Townsville and badly needed people to work under ground. I signed up there and then and made the 1000km drive out there in less than 2days.

In the late '60s, Mount Isa was a bit of a boom town. The 8-month strike was a memory. A lot of people who had been there before the strike left and never returned. New people were coming in all the time. There was plenty of work, and plenty of money to be made. Mount Isa was an affluent place and had everything that a town that size should have The problem was that it was basically desert country up there: nothing grew, and there was no dairy industry within 1500 km. Plenty of beef of course, but everything else had to be brought in at great expense. No fresh milk, just long-life. There were three fish 'n chip shops in Mount Isa. The fish was barramundi, caught fresh up in the gulf country, and flown down to Mount Isa for the fish 'n chip market. Best I've ever tasted!!

I stayed at Mount Isa for about six months, working underground. I settled into the work routine. It wasn't too bad. I even enjoyed it at times. It was tough work, but I was lucky I had a good workmate. He was an Israeli guy, and keen on earning money, as they are. I found it hard to keep up with him at times. Out of the three shifts on the contract, two were doing OK. But the third always seemed to be dragging the chain and were costing us money. My mate, the Israeli, was always leaving messages for them, which they always seemed to ignore. My bank was building up nicely, but I wasn't ready to move on yet. It was tough work at times, and my back was giving occasional twinges, but I kept going for the time being. I'd met some guys who were diamond drillers, and I was interested in doing that. There was also a lot of talk about West Australia, and Western Mining, and the nickel boom. Exploration was going on, especially in the Kalgoorlie region. There was a lot of diamond drilling over there.

Mount Isa to Kalgoorlie was a trip in excess of 3,000 kms, more than half of it on dirt roads, so we had to be well prepared.

It took 10 days including stops at Alice Springs and Ayers Rock and we arrived in Coolgardie in the afternoon. I didn't want to rush into Kalgoorlie today; I just want to stop somewhere, where I can get a shower, and freshen up, and make myself look a bit more respectable, before I go looking for a job.

The next day I had a bacon & egg roll and a cup of coffee for

Chapter 2 - Australia. The Early Years

breakfast and, feeling ready for anything, I took the short drive up to Kalgoorlie, and found the Western Mining offices. It was a new building, and fairly impressive. I went in, and discovered quickly that, yes, they need workers, and no, they don't need any diamond drillers at the moment, and was I interested in doing field work for them. OK, yes, I want a job, and I'm happy to start doing field work.

I became a member of one of the two IP teams, doing electronic surveys on land that a prospector thought suitable and where the surveyors had been through and established a grid. These areas were often remote and we would stay onsite for 2weeks at a time, living in caravans, coming back into Kalgoorlie every second weekend.

All that area, including Kalgoorlie, is desert, and daytime temperatures would regularly exceed 50C. Even at night, it wouldn't get much below 40C. Although it was desert, there was plenty of ti-tree scrub around, and spinifex grass (spear-grass, some people called it), so as soon as you got off the road, it was tough going for the vehicles. Western Mining was a company that had been around for a long time. They were a goldmining company, and they had survived the ups and downs for many years. Then along came the nickel boom, and gold was forgotten for the time being. When I was there, they already had a mine in production at Kambalda, and a number of other prospects. They had a policy of issuing shares to their employees, hoping no doubt that it would encourage people to stay with the company. I was pleasantly surprised when, after I'd been there for six months, I was given 1,000 shares. They were 5c shares worth about $1each at that time, but the potential was enormous, which I didn't fully comprehend at the time. If I'd held onto these shares, they could have set me up financially for life, but I sold them didn't I, and passed up a golden opportunity. I had been hearing was that the Ord River was getting ready to start up again. The wet season was coming to an end, and work on Stage 2 of the project would be starting up shortly. Stage 2 was the building of the second dam, which meant there would be lots of drilling and blasting going on. My experience at MIM would be valuable when it came to getting a job.

Down south, winter was starting to approach, and down there it could be cold and wet. My thoughts were turning that way anyway, when something happened. We'd finally completed the three large areas up near Leonora and were all packed up and ready to go on the Thursday. We were taking the caravan back to Kalgoorlie on Friday, but we had a few beers in Leonora on the Thursday night. Towing the caravan, it was a pretty slow trip back to Kalgoorlie, and were back in Kalgoorlie late in the afternoon. We put all the vehicles and the caravan into the yard, and walked back up to the single men's accommodation, which was provided by the company for a very nominal nightly fee. Please keep the place clean, was the only request.

As we neared the building, I could see people standing outside, and there was some kind of argument going on. I could see Terry, and Dave, the leader of the other IP team, having a discussion. We soon discovered what Terry had really come to see us about: The IP programme was being closed down over winter, and we were unemployed as of now. If we come to the office on Monday morning, we'll get the wages owing to us, and maybe apply for another job within the company. Those who wanted to stay at the accommodation could stay for the three nights if they paid now. That was fine by me. It was a lot better than sleeping in the car, and I could also have a shower, which was important.

As it turned out, Dave and I were the only ones to stay over the weekend, and by the time Monday morning came around we had a plan. But first… It was Saturday morning, and I was up early, then down Hannan st. to a café I knew, where I had a bacon and egg roll and a cup of coffee. Now it's opening time, and I'm across the road for the first today. The 'Exchange' pub is the first that you come to as you go down Hannan St from the single men's accommodation. It's not the best pub in town, but it has the best barmaid. I first met Carmel a few months previously. I'd

Chapter 2 - Australia. The Early Years

gone into the pub early one Saturday morning, and there was hardly anybody in the place, just me and another guy over the other side of the bar. The barmaid seemed to be in close conversation with him, and I was wondering when I was going to be served, when they both straightened up, and he left the pub, and went across the street and disappeared round the corner. 'Hi, I'm Carmel,' she said. 'What can I get you?'

'Just a pot (seven-ounce glass), thanks very much.'

She served me and walked away, and I couldn't help admiring her.

What an attractive gal she was, maybe 25, 26, 27, and so aware of herself. I watched her walk out of the pub and cross the road and round the corner where the bloke had gone. Oh, that's interesting, I thought. Another time, I was there early, and a similar thing happened, just a different guy this time. I hung around to see if she would come back. I had three or four more drinks, but she never reappeared. I asked the barman who served me, What's the story with Carmel? Is she on the game, or something? He stared at me for a moment, then he said, Carmel runs her own life, and she has her own friends. I wouldn't push her too hard if I were you, or you'll be sorry. Yeah, I hear what you say. I didn't go there the next time I was in town, but the time after that I thought I'd give it one more go, and so I walked into the bar a few minutes after opening time, and she was there on her own. She came over straight away.

'Hi,' she said, 'I hear you've been making enquiries after me.'

'Yeah,' I said. 'I was wondering if I could become one of your friends?' She stood back and laughed. 'I think I might be a bit expensive for you darlin'. You could go down to Rose St right now, no waiting.'

'It's not so much about money,' I said. 'It's about meeting an attractive sexy lady who turns me on.'

'What's your name,' she says.

'George,' I said. 'I'm not a local, as you know. I work for Western Mining, doing bush work, and I'm only in town every second weekend.'

'Just a lonely young Scotsman,' she said. 'Are you married?'

'No. I'm too much of a ratbag; nobody'll have me.'

'How much money have you got in that wallet of yours?' she says.

'A few hundred,' I said. 'I don't know exactly.'

We talked about price, and I agreed. An hour would be enough.

'OK,' she said. 'Just cross the street and go round the corner, then go down about 50 yards, and you'll see a green door.'

A green door? Reminds me of an old song, 'What goes on behind the green door'.

She laughed. 'Just knock on the door then go in,' she said. 'I'll be along shortly.'

So, I had a fling with Carmel, but it cost me. It cost me my Western Mining shares. When I went to the broker to sell them, he told me you're mad, man. These shares are going to take off any day soon, and you are going to miss out on thousands. I didn't care. I was in love with Carmel.

For the record, I was given the 1,000 shares by Western Mining. They were 5c shares, and at the time I sold them they were worth about $1.20. In the crazy months that followed, these 5c shares went to $50. The company then made a 10 for one issue, which brought the share price back to $5, then they took off again, and went right back to $50. The demand was still there, so the company made a 5 for 1 issue, reducing the share price to $10. It stabilised then around $10-$15 for some years. You don't have to be Einstein to work out that I could have been set up for life. Of course, I was just a young bloke, and enjoyed getting on the piss, and it is almost certain that I would have found a way to piss it up against a wall.

The stockbroker was right from his point of view. It would have been a radically different life if I had held on to them, but we'll never know, will we!

The plan that Dave and I had was to get out of Western Mining and out of Kalgoorlie. To head north to where the winters

Chapter 2 - Australia. The Early Years

were warm and find work. I had an open mind, but I was interested in getting up to the Ord River, where work on the second stage of the irrigation scheme was about to get under way. They were building a second dam, much bigger than the first, so that they could extend the irrigation area. Building a new dam meant there would be plenty of work for percussion drillers, and blasters, very similar to what I'd been doing in Mount Isa. I'd have a good chance of getting a job like that. Dave was happy to come with me anyway. He didn't have a car and would have to get the train back to Perth before starting all over again.

But first things first. I was due to see Carmel tomorrow morning, and I wasn't going to miss that. It would be the last time, but I would never forget Carmel. So, Monday morning, and we were at the Western Mining office; they'd paid us by cheque, and now we were going to have to go to the bank to get the cash — a small hold up; we had to pay a fee to get the cheque cashed. Finally, we're on our way. It's a trip of over 2,000 kms, so it'll take a while, all dirt roads. We might make a few stops along the way. The wet season is over now, so the roads should be open; hadn't heard anything to the contrary. Ora Banda, Menzies, Leonora, Wiluna. That's enough for one day.

The road between Wiluna and Meekatharra was poor quality, and we had to slow right down. We also had a puncture, which was fixed soon enough (with two experts on the job). On the way from Meekatharra to Port Hedland, we stopped to look around the town of Wittenoom. I could remember stories in the Sydney media about guys working at Wittenoom earning huge money. We were in the area now, so let's have a look. The mine is closed; the town is deserted. We found a shop that was still open and got the story. Asbestos is a dangerous material to work with. A large percentage of the people who worked here got asbestosis and died miserable, painful deaths. The mine has been closed, people are being forced to leave, and the road will be completely closed off.

We'd decided to stop in Port Hedland for a day or so It's a major port for the export of iron ore, and all the iron ore miners have offices here. So, we tripped around the offices making enquiries about jobs, and got the same story from all of them. We do not handle employment from this office. All employment is handled in Perth, and we can't offer you anything here. Dave was disappointed. I think he was regretting not going down to Perth from Kalgoorlie, but he decided to stick with me.

The next stage was from Port Hedland to Broome, a distance of over 400 kms. There was nothing in between, so we had to make sure that the car was in good shape, and we had enough petrol for the whole trip. That meant we left with a full tank and two ten-gallon jerries. We were into the second jerry by the time we got to Broome.

Nobody had said anything and there were no signs on the road, but we hadn't gone very far when we came to the De Grey River. There was no bridge, the river was running a torrent, and what had been the crossing was completely washed away. There must have been rain up in that area recently, and it was obvious that there wasn't going to be any crossing there any time soon. There was a railway bridge, a trestle construction, built for the iron ore companies. Could we get up there? 'You're mad,' said Dave. 'You're stark raving bonkers.'

'Let's walk over there and have a look,' I said. 'It's not far.'

As soon as we left the road, we could see the tracks of other vehicles heading over to the railway line. When we got to the line, we could see the wheel tracks going up onto the line and heading north. There wasn't a lot of room, but other vehicles bigger than us had done it, so…

'You're mad,' Dave said again. 'D'you wanna stay here,' I said, 'or walk along behind me?'

'A bloody train'll come along,' he said.

We could see a fair distance both ways — nothing coming; we had at least 15 minutes. We're up on the line, we have about three feet each side for the wheels, and we have clearance over the rails, and off we go, from sleeper to sleeper… Now we're over the river, and there is nothing between the sleepers. Don't look down for Chrissake!

Chapter 2 - Australia. The Early Years

We make it across the river, but still a fair way to go before we can get off. We find the place where other vehicles got off. It looks a bit steep. We're sliding down the bank; we nearly roll over. Turn the wheel, just a little... We're through!

Back on the road, we stop for a breather. It had taken us 20 minutes.

'You're bloody mad,' says Dave, repeating himself again.

The rest of the drive was fairly straightforward, It's just the distance. We had to stop about halfway, and fill up the tank from a jerry can, and again before we got to Broome; 600 kms, and not a sign of habitation, not even a side road with a signpost.

Broome has been a settlement for a long time. It was the centre of the pearl fishing industry. Now that is mostly gone, but it has become a tourist attraction, and tourists fly in every day. We spent a couple of days there, getting ready for the next leg. If you leave Broome today and travel to Kununurra, there is a nice, sealed road that crosses a bridge over the Fitzroy River and goes into Derby, then on to Fitzroy Crossing, where another bridge crosses the Fitzroy Fiver, and the road then goes on through Halls Creek, etc. It's now well over 50 years since we did the trip, and I am almost 100 per cent certain that these bridges did not exist in 1968.

I am almost certain that we did not visit Derby, or Fitzroy Crossing, but instead we travelled down a bush road that kept us south of the river, and we did no crossings. That track is still marked on road maps today. What I do remember for sure is that, when we got to Halls Creek, we became minor celebrities. 'Where have you come from?', everybody wanted to know. Dave was happy to tell them all the story of the railway bridge. I thought it was all a bit of a dry argument, and headed for the pub. 'Sorry,' the barman said, we ran out of beer three weeks ago.

'Would you like a shot of rum?'

'When's the beer truck going to arrive?'

'We don't know... could be weeks.'

We rested up at Halls Creek for one night. That was enough. Next day, as we were getting ready to leave, the local policeman came to see us. 'You're going up to Kununurra?' he asked us. We said yes.

'You've got 12 creeks to cross,' he said. 'People are going through in 4wheel-drives, so you'll probably get through. Just take it easy.'

The wet season was well and truly over, but the streams and rivers continue to run for at least another month, and we'd misjudged that. We got some extra petrol in Halls Creek, and it was just as well we did. The 300 kms trip took us two days.

Kununurra was a new settlement, built specially for the workforce required for the dam building. There was a shop, a post office, and a company office. A notice on the door said, 'Staff will be here in June'. There were a few guys hanging around, waiting for the work to start up, and it seemed like they already had jobs fixed up. The employment office was in Darwin, they said. That's where you gotta go.

There was no pub, and the shop had nothing but sugary soft drinks. We might have to go down to Wyndham Dave, I said. 'What do you reckon?'

'Let's go,' he said.

Wyndham was established as a port in the goldrush days, and it seemed like a fairly run-down place when we were there, but it had a pub, and there was plenty of beer. We had a few, and a bit of a chat with some of the locals. The news was work wouldn't start until June at the earliest, and maybe we'd get a job and maybe we wouldn't. Best chance would be to go up to Darwin and go to the employment office. I knew that Dave wanted to get to Darwin, and maybe go his own way from there.

That was OK. It was probably the best thing for me to do, too.

There was just one thing. We'd have to cross the Victoria River, and the main crossing has been blocked by a semi-trailer which had been stuck there for three weeks. There was another crossing a couple of miles farther up the river. 'You should be right there, mate,' one of the locals said to us. 'Just take it easy.' Thanks for the advice.

Chapter 2 - Australia. The Early Years

Sure enough, we get to the Victoria River, and there is the semi-trailer, and a disconsolate driver sitting at the side of the road. We stopped and had a chat with him. He was OK. He had food and water and was expecting a tow truck any day now. 'Don't try and cross here,' he said. He echoed the bloke earlier. 'There's another crossing further up. You'll be right.'

'Have you seen any crocodiles?' we asked him.

'Nah; no salties this far up,' he said.

We found the second crossing, and it looked OK from where we were, but I wanted to walk across it to check it out. It was about 100 m wide, and very shallow at our side, but as you went across it got slightly deeper, and it was running faster. We were probably OK with that, but there was a bank to contend with. It had been broken away with the passage of vehicles, but it was definitely a bit rough. Could we get up there? We spent a bit of time breaking it down a bit more, to give ourselves a better chance .It's getting on in the afternoon, and we don't want to spend the night here. Back in the car, put it in low gear, and in we go. Steady as she goes. We pass the half-way mark, everything OK; but now the water is a little deeper, and it is starting to come up the door. We're nearly there, but the water is higher, and it could push us sideways. The front wheels have reached the other side, and we're starting to climb up the bank. We'd put a whole lot of dry rubbish on the wheel tracks to try and give us more traction. It worked, and we drive right up and onto the road. Phew!!!, we made it.

'You're mad,' says Dave.

When we arrived in Katherine, my watch was saying 8.30 pm. Just time for a couple of beers before closing time. Where's the pub? We didn't have any trouble finding it. There was a mob of drunken Aboriginals outside it. The pub was closed. Then the penny dropped. My watch was on West Australian time, and we'd crossed the border into the Northern Territory, where the time was 10pm. We tried to plead with the staff, but no dice; the Aboriginals were giving them too much trouble. 'Not tonight,' they said. 'Sorry!'

Next day, we had an easy drive into Darwin on a sealed road, and easily found a pub that served counter lunches. After lunch, we followed directions from the barman and drove to the Ord River Project office. The news there was disappointing. They wouldn't be ready to start for a couple of months, and they weren't employing people right now. Go in and see the people next door, they suggested. They're looking for people. 'Western Nuclear ', the sign said, as we walked onto the property, and when we made enquiries they said, 'Yes, we need people.'

Western Nuclear was a Canadian company in uranium exploration. They were operating in two areas, the general area of the Mary Kathleen mine in North Queensland, and in Arnhem Land in the Northern Territory. In Arnhem Land, land transport is very difficult, and they operated mainly out of helicopters, whereas in North Queensland they are based in Cloncurry, and use 4WD vehicles. It was decided Dave was going to Arnhem land, and I was going to go to Cloncurry. Dave and I said goodbye and went our separate ways.

I felt that Cloncurry was a big step along the way back to Sydney, but there was no hurry. It was more than a day's drive, and I wanted to time my arrival in Cloncurry for mid-to-late morning. I didn't stop in Mount Isa on the way through and arrived in Cloncurry in mid-morning. The company operated out of a rented house in Ramsay St. I met the boss, a Canadian guy, and three geologists (I can't remember their names).

All the field hands were out working, but I met them all later. The deal was, pay was not all that high, but we all lived in the house, and had all our meals covered — breakfast lunch and dinner at the Post Office Hotel, the best pub in town. You had to buy your own booze of course (that went without saying).

The work was simple enough. We were working in and around the Selwyn ranges near the Mary Kathleen mine, walking up dry creek beds, picking up samples at regular intervals, each sample into a small, brown paper bag with identification, and

Chapter 2 - Australia. The Early Years

stowed away, to be analysed by a geologist later. We always worked in pairs, for security reasons. Sometimes, we came home at the end of the day's work, and sometimes we stayed out for a week depending how far away we were.

There was a pool hall not far up the street from our place, and some of us would go there after the evening meal. It was owned and run by a guy called Bob Backash. Bob had Afghani heritage, and his family went back to the days before the railways were built; the days when camels did a large percentage of transport in the hotter areas of Australia. Camels were better than horses in these parts, and thousands were imported. With the camels came the cameleers, mostly Afghani, and today it's not uncommon to find people whose ancestors go back to that time.

Bob Backash was also a bookie and fielded at all the race meetings around the area, especially Mount Isa. Horses racing in Australia all have a medical check by the vets on the morning of race day, sometimes known as the 'blood count'. The information that comes out of the blood count is supposed to be confidential, but when you consider that that information can give a strong indication that a particular horse is primed to win, vested interests, such as bookies, make sure that they are in on it. So, on Saturday mornings the phones are running hot all over the country as the news goes out. One of my workmates, Ronnie Hamilton, was from Sydney. He was a keen punter, and knew all about the 'blood count', so on Saturday mornings he was always down at the poolhall talking to Bob.

'Waddaya know, Bob?... Come on, give us something.' And after a bit of pestering, Bob would give us a couple of horses' names racing in the southern states. There was a TAB in Cloncurry, but a guy like Ronnie liked to go to the races to be in the middle of the action. His problem was that he didn't have a car. I had a car, and I was getting interested in the horses, and it wasn't long before Ronnie and I were going up to Mount Isa every Saturday to the races.

Bob Backash's tips were good for us. We didn't always win; probably a bit more than half of them were winners; and you could also have a saver, so if it didn't win at least, you'd get your money back, plus. BobBackash would be there himself on his stand every Saturday taking bets, but we were told, we don't bet with him, or there'd be no more tips.

One day, we'd both been successful and had big days. It was nearly the last race, and I was looking for Ronnie as it was nearly time to go home. I found him over near the Marshall Boyd stand. He was talking to the man himself. 'Hey man,' he says, 'we're not going home straight away.

We're invited to the Boyd Hotel.

'Yes,' says the man himself. 'Come down to the hotel for the night and be my guest. The girls will look after you.'

As soon as we got away from Boyd, I asked Ronnie, 'How do you know him?'

'I knew him a few years ago,' he said. 'I was a strapper at one of the major stables (in Sydney), and I used to pass on information to him. He's not such a bad bloke really; he recognised me after all these years. 'These gals' cost $500 a night each,' I said. 'That's some kind of favour.' 'Yeah, but we gotta sling the gals, don't forget,' says Ronnie.

'Just as well I've had a good day then,' I said.

It was about 1am, and we were cruising along the road back to

Cloncurry. It was a good road, sealed, mostly straight, and you'd do the 80 miles in about an hour at this time of day. We had just passed the Mary Kathleen exit, and we were travelling nicely, then there were the taillights of a car in front of us. The car was stopped, and the driver was out on the road, staggering around as if he was drunk. I stopped our car behind him, and we got out to see what was wrong. It was obvious straight away: there'd been an animal on the road, and he'd run into it, full pelt. The animal was still alive but would have to be put down, and the front end of the car was completely smashed. The driver had cuts on his head, where he had hit the windscreen (no seatbelts in those days).

'What happened mate?' I said, "Did it jump out in front of

Chapter 2 - Australia. The Early Years

you.' As I was talking, I suddenly realised who the guy was. It was the copper from Mount Isa. The detective sergeant. Oh shit! I wasn't exactly sober myself. He was talking to Ronnie. Could you guys give me a lift into Cloncurry? he was saying. We hesitated for a second. He pulled out his police badge. We need to go now he said. We dropped him off at the police station, which was just round the corner from our house. Don't say anything to anybody, we agreed; just shut up, and say nothing.

Across the road from us was the dump of a hotel, the Prince of Wales. We didn't go there very often; it wasn't a very welcoming place. The licensee was a fellow named William Swan, or Swanee to you and me. He'd get up early in the morning, and get down into the cellar, and fill up stubby bottles with draught beer. The Abos were his main customers for these, and they'd buy him out every day. Of course, he'd sample one or two along the way, and by opening time he'd be half cut. Swanee had a barman, an ex-boxer who looked the part. He was a bit loud and aggressive and hated the Abos. What happened to him in the end was almost inevitable. It happened on a weeknight. We'd been to the Post Office hotel for our evening meal, as usual, then repaired to the pool hall, where Bob would have a few beers in the fridge for us, and we played snooker for an hour or so.

About 9.30, Bob came rushing in. Something's happened at the PoW, guys. The coppers are there, and ambulances. Of course, we all rushed out to see what had happened. The street was blocked off and we couldn't even get to our place. Our boss came out and spoke to the coppers, and we were allowed through. What had happened was predictable. A bit after 8pm, when everything was quiet two Aboriginals had walked into the pub and wanted to buy some stubbies. There was an exchange between them and the barman, who refused to sell them beer and kicked them out. Not long after, one of them comes back, walks into the pub with his sawn-off shotgun, and gives the barman both barrels from short range.

When the dust settled on that incident, a barmaid appeared at the PoW. This was better. Everybody wanted to meet the barmaid, and Swanee had some customers again. Her name was Carol, and she was a nurse.

She had worked at many of the church missions around northern

Australia. She had a nice personality and was always on for a chat.

A few weeks after she came there was a buzz around the place, one of the big social events of the year was coming up. One of the big cattle stations was having its annual picnic race day. There was a lot of talk about who was going to go. It sounded like a big day. On the Saturday morning of the event, we were supposed to meet outside the pub. I only had to walk out the door of our place, and across the street. There was nobody there. I stood there for a couple of minutes, but nobody appeared. I was just about to give it away when Carol came out of the pub all dressed up and looking smart. Looks like just you and me, she said. That's fine I replied, let's go and have a nice day out.

The cattle station was Clonagh, and it was about 80 kms north of Cloncurry along a dirt road. Carol knew where to go, and it didn't take too long to get there. The weather was hot and dry, which it is at that time of year. There was a good crowd there, and the races were about to start.

The situation was the same as Mount Isa, where the bookies will bet on the local races, but the main action is on the big city races. Bob Backash had given me a couple of tips, which was the basis of my day, and I had another couple of wins on my own account, which meant that I'd had a good day. After the racing, there was a huge barbecue, which went on for over an hour. Carol knew lots of people there, and she was roped into helping with the drinks. After a lot of people, left she came over and joined me, and we had our barbecue together. Music was playing somewhere, and Carol suggested that we have a couple of dances before we go home. A barn had been prepared as a dance hall, a few tables and chairs, music, a bit of sliperine on the floor, and away you go. We found a table and sat down for a moment. The music started up again, and we got up to dance. I hadn't danced

Chapter 2 - Australia. The Early Years

for years, it felt strange, but nice.

'Did you enjoy your day?' I asked her.

'Yes,' she said. 'Will we have a nice evening?' A small squeeze of the hand. Hmmm, I thought, I might be onto something here. Then a bloke who was there walked right across the dance floor, right across to our table. He picked up my glass of beer and started drinking out of it. 'Hey, excuse me,' I shouted, and I moved towards him. He turned slightly and threw the remaining beer in my face. Next thing I knew, he'd hit me with an uppercut to the chin. I saw stars, and down I went. Next thing I knew I was being kicked by at least three people. I tried to struggle to my feet, and all of a sudden, they were gone.

I was in shock. No police here; no security of any kind. Who were they? Carol was there. She knew one of them, 'but never mind that for now,' she said. 'You need to go to hospital and get stitches in that eye of yours.'

I told her I was ok.

'Huh!' she said. 'You men, you're always OK aren't you. Give me the car keys; I'm driving, and you're going to hospital.'

We talked on the way in. She said, 'You were winning too much. Everybody wanted to know, Who's the Pommy bastard winning all the money? And then the last race, some of them didn't like that.

'They're a wild bunch up here. The family think they're the lords of creation.'

I hadn't been going all that well, just a little in front up until the last race. In the last, there was a mare running that I'd been watching for a few weeks. She'd come back from a spell and had run in a couple of shorter races, which didn't suit her. This time, she was in a race against moderate opposition at a distance that suited her much better. I thought, she might have been set for this race, and I'm going to have a go at it. Her name was Grey Gull – a little memory from 50 years ago – and I had $20 each way at 12/1. She got clear in the straight and won going away. That meant that I'd had a big day and was probably what got up the noses of the wrong people.

You're going to meet Dr Harvey Sutton; Carol was telling me. I'd heard of Dr Harvey Sutton. He was famous in this part of North Queensland for his service to the community, which was well above and beyond. When he retired, he was awarded an OBE, and ABC television did an Australian story program on him.

It was late when we got to Cloncurry hospital, and he wasn't there, but he came in about half an hour later. Who's done this to you, he wanted to know, and he named a name. It wasn't him; it was his younger brother, Carol said. 'Oh, him,' said the good doctor. 'He'll get his comeuppance one of these days.'

My eye was the worst. The brow was split open and bleeding. He cleaned and stitched it and put a plaster over it. My nose had taken a belt, too, and was a bit out of shape. 'Can't do anything about that now,' he said. 'You'll have to live with that.'

It wasn't far from the hospital to our place. I was trying to thank Carol for sticking with me and looking after me, as the car drew to a stop. I started to open the door, but she grabbed my arm. 'George,' she said, 'you've had a tough day, but don't go running out on me now. Your day isn't finished yet!' Ok.

I snuck back across the road around dawn hoping to avoid everybody, but Bill, the head geologist, was up and about. He was aghast, and I had to tell him the story. You won't be working this week, he said. I'll be right mate, I'll be right, I said. I'd be there. Then the other guys were getting up, and we were going up to the Post Office Hotel for breakfast. Boots Schumacher, the manager, was there, and he'd already heard about the incident. Have you reported it to the police, he wanted to know? 'No, I haven't, and I'm not going to,' I said. 'I'll just put it down to experience.'

'That whole family are a bunch of ratbags,' he said, 'and they're barred from this pub. You'll never see them here.'

I told him I was glad to hear it. The Hotel had been in Boots' family for generations; his mother was the current licensee, while Boots was the manager. He was a friendly sort of a bloke, but very businesslike. He wasn't a big guy, but he had a big beer belly, and

Chapter 2 - Australia. The Early Years

when he encountered a problem customer, he would belly bounce them out the door.

I didn't go back across the road to the PoW for a few days, to let the dust settle, and when I did, I discovered that Carol had gone. Swanee wouldn't or couldn't tell me how or why she'd gone, but I had a chat with Boots one day, and he told me, 'Yeah, she's a funny one, that. She's got friends and family all over the north, but she won't settle. Spent too long in the missions. They were dreadful places; everybody knows that.

There were tragedies in her past.'

I thought she was a very nice person, I said.

'Yeah, everybody thinks that' he said. Do you want to contact her?' I told him no; she'd made her choice; that was fine by me.

The next Saturday, Ronnie and I went up to the Mount Isa races. We had our couple of tips from Bob Backash, as usual. One of them was in an early race, and it ran a place, so I was more or less square with that one.

Then I had three other bets that lost, so I wasn't travelling too well. Bob's last tip was in the last in Melbourne, and I was sitting out on the grass waiting for the race to come up. There was a bloke standing about 30 feet away glancing in my direction. He started walking towards me. That's the bloke from the other night. What does he want? The guy walked right up to me and said, 'I'm the brother of the bloke who gave you a touching up the other day'. Oh really!

'I just want to apologise on behalf of our family,' he said. 'He was out of order, and that was totally unnecessary.' Yeah!

'Did you report it to the police?' No!

'There are people around here who think you did.'

'No, I didn't go to the coppers,' I said. 'I just put it down to experience. But tell me something. I had $800 in my wallet that night, and they never touched it... so, what was it all about? 'He said, 'You're a stranger, see, and making too many winning bets, they notice that, and they don't like it, and then you were with Carol...' 'Carol?' I said. 'How does Carol get into it?'

'Carol's one of us he said, and we don't like it when she goes with strangers, especially Pommy bastards.'

He was starting to get a bit revved up, and I thought I should get out of there.

'Look man,' I said. 'I have no problem with you, and I wish you well. This is a public place, and I'm as entitled as anybody to come here and have a few bets. See ya later.' And I walked back into the betting ring.

They were betting on the race that I was interested in, and I checked around the boards. They were offering 7/1 about the horse I was interested in. I watched them for a minute or so, and one guy changed his board to 15/2. I claimed him straight away for $20 each way. Matey was watching me; I could feel his eyes boring into my back. It was getting close to the start of the race, and I saw another bookie put up 8/1. I claimed him, just in the nick of time. The race was close, but we won in a photo finish. Yippee!! I'd had $80 on the race, which was a lot for 1969, but it was Bob Backash's tip of the day, and I felt that the worst that could happen would be that it would run a place.

I finished over $300 in front on the day, which was excellent. It was just a little worrying to me that people might be noticing our success, and the wrong people at that. I had a chat to Ronnie on the way back to Cloncurry. Don't worry about it, was his advice. A few more weeks, and we'll be closing down for Christmas, and we'll all be out of here.

'You shouldn't have gone to Clonagh,' he said. 'No security at these places. Anything can happen.'

Next morning, at the house, there was a big announcement. We're closing down this operation in three weeks' time. There is a lot of work to be done in three weeks, so you'll be going out, and staying out till Friday night, then back out on Monday, etc. I was paired with Albert, a local Aboriginal guy. We were going to be living in tents, and we had a whole bunch of gear to put together, and food supplies for the week.

The weather was hot and would remain hot until the following May. The plan was to work early, then knock off about

Chapter 2 - Australia. The Early Years

10 am, rest up during the heat of the day, then go back in the late afternoon to do a few hours when it was cooler.

We were going to the Selwyn Ranges around Mary Kathleen, and the dry beds of the streams that ran off the ranges in the wet season. We were given maps. Albert had intimate knowledge of the area, being a local, and having done this kind of work before. I was lucky with Albert in other ways. He was just an all-round good guy. He was always on for a chat, and he cooked the best steak you ever tasted.

It was a tough week. The heat was severe, and I struggled to keep up with Albert. On the Friday morning, we went out early; there was one more stream bed in my little area, and I would get that done and we would pack up for the week and get back into town. I was most of the way there, and the land was rising steeply, and the stream bed was getting harder to follow. Up above me, I could see a large, flat rock jutting out. I'll get to there, up on that rock, I thought, and I'll be able to have a good look around and see what's what. So, I clambered up, the ground getting steeper all the while. Finally, I reached the rock, and started to straighten up. My eyes just reached the level of the rock, and Whoaaa!!! Lying on the rock was a huge rock python sunning itself. It must have heard me struggling up the hill as its head was turned towards me, not three feet away from my face, its forked tongue flicking in and out of its mouth. I took off like a scalded cat.

How I never fell over and injured myself, I'll never know. When I got back to the camp, Albert was already there. He had a good laugh when I told him what had happened. It was probably still digesting its last meal, he said, having a nice sleep in the sun when you came along and disturbed it. There are plenty of snakes that would have attacked you in that situation, but not a python. That was most unlikely.

We weren't going to do any more that day. It was time to knock off. We got all the gear together and headed into town. I had a shower and a shave, and put some fresh clothes on, and felt much better. Ron was there, and we talked about the races the next day. The Melbourne Cup carnival was over, but there was still good value to be had. We'd see Bob Backash in the morning.

The next day, Ron and I went up to Mount Isa to the races, as usual. It was a fairly average day, and I was lucky to finish marginally in front. Ron didn't want to stay late in Mount Isa, which was fine by me. On the way back to Cloncurry, I spoke to him about going back to Sydney; maybe he could come with me and share expenses. No, he said, he wasgoing down the coast to visit family, and then down to Sydney in the New Year.

Ron was a bit of a dark horse. He knew the racing industry inside out and had been a strapper back in the days when Marshall Boyd was fielding on the rails in Sydney. Strappers were very often party to confidential information, and he may well have had a relationship with Boyd. I noticed that every time we went to the races, they had close conversations. Bob Backash , too, had a lot of time for Ron, and had long conversations with him. He gave Ron good tips, which I don't think he would have given to me. I had a plan to become a punter when I got back to Sydney, but it wouldn't be the same without Bob.

When we got back to the house, everybody was there milling around. We had to unload the truck, and by the time that was done, Bruce (the boss) was ready to address us. He waffled on about it being a good year; how all the targets had been achieved; and now it was the end of the year, and the budget was running out. Somebody asked him whether they'd be back next year.

'Only if it's in the budget, he said. 'It's not a high priority. If we come back, it won't be until the winter.'

So, we all got our cheques, and then he made one more announcement: the local guys were free to go, 'All the best for Christmas,' he said to them. 'To the guys staying at the house, you don't have to leave this instant; you can stay for a day or two if you wish. You may go to the hotel for breakfast but no other meals, except for tonight.'

I went looking for Ron. What do you reckon I asked him? The races tomorrow, he replied. Beauty, I said.I got a good night's

Chapter 2 - Australia. The Early Years

sleep, and then they were getting up for breakfast. Ron was in a good mood as we headed up the road to Mount Isa.

'Bob's given me a couple of goodies today,' he said. 'We could be having a big day today.' I knew exactly what was on his mind. I fully intended to bet big on Bob's tips today; this would be the last time that I'd be at Mount Isa, and I wanted to make the most of it. One of Bob's tips won, and the other came third. That was a good start. Then I had five other bets, where two of them won, and another came a place. That meant that I had cleared over $1,000 on the day; a big, big day.

Ron similarly had had a big day and he was grinning from ear to ear when I caught up with him. Winners are grinners, I said to him. 'Yeah,' he said, 'we're booked for the Boyd Hotel, OK?'

'Yeah, you bet, I said. 'This will be the last time, so let's make the most of it.'

The sun was up, and people in the house were getting up, and going to breakfast by the time we got back to Cloncurry. A quick shower: the rest of the day would be for sleeping and resting up, I had a long drive ahead of me, starting on Monday morning.

The previous week, I'd tried to ring George Howie at work. I was hoping I could stay at his place, just for a few days till I organised myself. He wasn't available, but the telephonist, whom I knew, asked me to leave a message, to ring back on the day I left Cloncurry, and she'd have an answer for me.

The route I was taking was inland, which was much shorter, with less traffic, than going back to Townsville.

When I got to Winton, I found the Post Office and made the call to

Sydney. Sylvia (the telephonist) recognised my voice straight away. George, she said, I have a message for you. Do you know Armstrong St in Ashfield. Yes, I think so. 'OK, it's number 1, the bottom flat. The key will be left under a stone near the door. There'll be nobody there, just go in and make yourself comfortable, and we'll contact you later.

'I'm expecting to be there late Thursday,' I said.

'That's OK, she said. 'We'll probably contact you over the weekend.'

'Who's "we",' I said. 'Is that the "Royal We"?'

'You'll find out,' she said, with a laugh in her voice. So, goodbye to Cloncurry, and Mount Isa; what an amazing few months that had been. I felt like a different person.

I'd probably left there at just the right time. It couldn't have gone on much longer. Sydney was beckoning me. The trip took a bit over three days, mainly dirt roads in Queensland, severe heat during the day, not a lot of rest. As I got closer to Sydney, I started to think about what I was going to do. I was going to start off by going to the races. How would I go? There'd be no Bob Backash there to give me hot tips.

I found the address in Ashfield without trouble, and found the key, and moved in. A shower and rest were high on the agenda, and it was well into Friday morning before I surfaced. Hunger drove me up the street, and I got the paper while I was there. The races were at Rosehill this week, and I decided to go. It would be my first experience of Sydney races. I tried to ring George Howie in the afternoon, and this time I got through. We had a bit of a chat, and he told me that he and Sylvia, had moved in together, and I could stay in the flat. That was fine by me, and we made arrangements for paying the rent. In the longer term the block was going to be knocked down and redeveloped. I could understand that the place wasn't in great condition, but it suited me fine for the moment. How about a few beers? I thought you'd never ask. How about the Summer Hill pub, lunchtime tomorrow,

I'd been to Summer Hill a few times, before I went away, and I knew most of the guys. There were Scots and English, in a good crowd of guys, and Saturday lunchtime was the time, everybody was there. The leader of the push was Andy Brough, a gruff Fifer. He was a war hero, and everybody looked up to him. He had joined the Royal Navy in the 1930s and was in the Far East on the Prince of Wales at the outbreak of World War 2. When

Chapter 2 - Australia. The Early Years

Prince of Wales and Repulse left Singapore to take on the Japanese, Andy was left behind; he had gotten some relatively minor injury and was unfit for duty. It was lucky for him that he was left behind, as both ships were sunk with total loss of life. As the Japanese neared Singapore, a group of Navy guys commandeered a vessel and took off, trying to get to Australia. After many adventures and help from the fuzzy wuzzy angels the few that were left were picked up by an Australian vessel and taken back to Darwin. When Andy recovered from his wounds, he was seconded to the Royal Australian Navy, and spent the rest of the war on Australian ships in the Pacific. After the war, one of the survivors of the escape wrote a book about their experiences, and later a movie was made. I was lucky enough to get a copy of the book, and read about their adventures, but I never saw the movie. Andy stayed in Australia after the war, and met and married Bonney (Yvonne), and became a member of the notorious Painter and Dockers union, which dominated the waterfront in the years after the end of the war.

Lunchtime extended into the afternoon, and before we knew it was early evening and time to go home. George had two messages for me. He was being posted to Melbourne for a while, and then on to Perth, so I wouldn't be seeing much of him for a while. And he'd been in contact with my old boss at Arnott's, Benny Simpson. Benny wanted to get in contact with me re coming back to work at Arnott's. Could I give Benny a ring to arrange a meeting? Wow I wasn't expecting that. I didn't want to go back to Arnott's or any other factory/office job. It surprised me that they might actually want me back.

I was going to have to do the right thing and give Benny a ring. I liked Benny, and the folks in the office, but what was I going to say to them? I thought about it for a few days before making the call. After exchanging niceties, I said that I'd enjoyed my time at Arnott's, and pass on my regards to the folks, but my world has moved on, and, thanks for the interest, but I wouldn't be coming back to Arnott's. I was in Sydney for the summer season and was due to go back up the bush in February. A few more niceties, and we left it at that.

I started going to the races, as I'd planned. This wasn't Mount Isa; this was the big time. I always went to the paddock, as Ron had advised me to do. The Paddock had access to the rail's bookies, and there was a holding area where the public could go and watch the horses parading before the next race. I was fairly cautious to start with, but I did win on the day three weeks in a row. This gave me confidence. I wasn't winning a lot, but I was winning. My system was working up to a point, anyway. I started to go to the midweek meetings, where the horses were a bit lower class, and the system didn't work as well.

I was playing golf twice a week, and spending time travelling around looking at boats. I'd be at the Summer Hill pub most nights and was starting to drink too much again. I went on till about early February; I wasn't doing too badly at the races, and had had a few coups, but my bank was going down. Living in Sydney was an expensive business, and I just wasn't winning enough to cover that. One bad Wednesday, where I lost more than I should have, was the clincher. I'm going to have to get a job.

I started looking in The Sydney Morning Herald, and it was only a few days later that my eye caught this ad, 'Experienced bush worker required', with a phone number. I rang the number, and it was an agency. They asked me to come in for an interview, which I did. The job was working for a company involved in the oil industry, working on oil rigs, and other places in remote areas. Could I handle that? Sure, no problem. OK then, we're going to send you to see the boss of this company called Data Analysis. They are just across the street at 261 George St. And that was how I came to meet Mike Gahan.

Mike was an oilman, and for a large part of his life had worked for Schlumberger, a company that provided specialist services to the oil industry. He then left, and set up his own company 'Data Analysis', which collected information from the drilling of oilwells and geological surveys, analysed it, and sold the results to mining companies. To that end, he employed four

Chapter 2 - Australia. The Early Years

geologists and two field hands, one of whom was me, to go out and get the information.

During my time with Data Analysis, I sat on one onshore well in Victoria, for Woodside; four wells in the Simpson desert for Pexa oil; four small wells in the Shoalhaven area of NSW for Woodside; one wildcat well in the Rolleston area of Queensland, which I think was Mike's own money, and another wildcat well in the Thompson River area south of Longreach Queensland, which was also Mike's money. Then there were two geological surveys, one in the north-west of Western Australia, where we were based in Broome, and another in the Gloucester area of northern NSW.

My first job was in Victoria. Woodside had drilled a series of wildcat wells in onshore locations in the Sale area of Gippsland, and this was to be the last. We'd been told that the Woodside top brass were coming down from Melbourne and were going to shout everybody a night out, and we were not going to miss that. We all gathered at a large hotel in

Sale and had a few drinks in the bar before proceeding to the restaurant.

We all got seated and settled down, and the chairman got up and made a speech, thanking us all for our efforts, etc. He got to the end, and there was a round of applause. 'Thanks very much,' he said. 'Before I sit down, and before all you guys get too drunk, I want to give you all a tip. He continued, 'Woodside is a very small company by even Australian standards, but that is about to change. We have found something very, very big on the North-West Shelf. It has not been officially announced yet, but it soon will be. When that happens, the share price is going to take off, so my tip is get in now with everything you've got, and you will never regret it.

'Thank you, guys, and enjoy your evening.'

Well, what do you make of that? I had nearly $5,000 left in my 'bank', and Woodside shares were 5 cents each at the time. How many shares could I buy with $5,000? I'm going to be right on to that as soon as I get back to Sydney! Then I discover that I'm not going back to Sydney straight away. One of the geologists, plus me, are going down to Melbourne, then flying to Adelaide, then up to the Simpson Desert where Pexa Oil were about to start their first well.

The pilot knows where the rig is situated, and there it is. A temporary airstrip has been prepared between the sand dunes, and soon we are on the ground. What a difference to cold, wet and muddy Victoria. Even though it's winter, the afternoon sun is hot, and everything is covered in dry, dusty sand.

The drilling rig here is much bigger that in Victoria. The drilling company is Richter Bawden, a Canadian outfit, who have a base at Roma in central Queensland. All their staff fly in from there.

There is a short-wave radio set up in the shed, and we get calls on it at least once a day, from Charlie Jones, a Texan. We reckon that Charlie is Mr Pexa himself. We also reckon that he is playing the stock market. Pexa shares are generally around 2c, but occasionally they issue a drilling report, and the shares shoot up to 4c or 5c. It's easy to play the market when you have inside information like Charlie does. So, the drilling proceeds, and we get a few 'shows' in the sand, and shale areas, but it is not a major find, and Charlie is not best pleased. The well is going to be finished shortly, and nothing has been found. A couple of days later we hit the basalt, and that is the end of it.

Back in Sydney, I had to report back to Mike, and find out what the next move was. I rang him in the morning, and he said could you come in and see us. The next Pexa well is not starting for three weeks, and we have another job for you in the meantime. I decided to get the train in as the office was right at Wynyard station. We're going to Western Australia.

Mike has received a request to survey a certain tract of land up in the iron producing area, and a geologist + me + gear, were going up there to do the work, which would take a week to ten days. The flights had already been confirmed, and we had to get a move on, and get the gear organised, and be at the airport in the morning. We flew to Perth, then caught a second flight up to

Chapter 2 - Australia. The Early Years

Broome. A vehicle had been booked for us, and there it was, just as well with all the gear we had. We loaded it up and drove to our motel. It had been a long day.

The next ten days were hard. Even though it was winter the temperature was above 30 degrees Celsius every day. The geologist wanted his samples from at least one foot below ground level, and that meant using the augur drill.

On the way back to Sydney we had all the gear plus a pile of samples. What that trip must have cost to collect this pile of samples, I just couldn't imagine.

When we got to Sydney, we split up. The geologist got all the gear into one cab and went his way. I got all the samples into another and took them back to Ashfield. It was late in the evening, and I'd ring Mike in the morning and see what he wanted me to do with them. I was to take them to an address in Kent St in the city, where Data Analysis had a lab. And so, I came to meet John, who was a chemist, and ran this lab on his own.

I am writing this in 2021, which is 50 years-plus since these events took place, and so it is not surprising that I have forgotten many or most of the names of people I met and worked with. John was one name that stayed with me. He was a battler, a family man, who had studied part time to get his chemical qualifications, and here he was stuck in this lab all on his own. I was a regular visitor from then on, and he was always on for a chat.

After I'd dropped the stuff off, I had to go up to 261 George St to see Mike about the next job. I was going to be on my own this time, except that there was going to be a geologist there from another company. I was to meet him in Adelaide, and we would go together on the charter plane up into the desert. The gear was already there, so all I had to carry was some personal stuff. The guy's name was Ned Hale, a Texan, who worked for an American company called Geoscience. He'd been brought in because Data Analysis didn't have anybody available to do it.

Ned was one of the good guys. I took to him straight away, and I
learned a lot from him. He seemed to be just the right guy to

deal with Charlie Jones, and they got along well. The well was completed without incident, nothing was found except for a couple of minor shows. Back in Sydney, Mike had another job for me. Down around the Shoalhaven area of the South Coast, Woodside had a small drill rig drilling wildcat wells. Could I go down there, and work with them, find out what they were doing, and get a few samples? So, I went down there, booked into a motel in Nowra, and went looking for them. What I found was a bit of a shambles. There seemed to be no-one in charge. The driller, who seemed to be the leader, had an attitude problem, and a drink problem.

They were playing up and got kicked out of their motel.

I had to ring Mike and let him know what was going on. He asked me to stay and try to get something out of it. He was going to ring Woodside and tell them what had happened. A couple of days later, when we started up again, the driller had disappeared, and had been replaced, by a much more sensible character who went out of his way to help me do my job. So, the program was finished, and I got all the samples I required for three out of the four wells, and I was very happy to get back to Sydney in time for the next Pexa well. I hadn't forgotten about Woodside shares, but I'd been very busy, and had never had a chance to go and take up the offer. So, after I got back to Sydney, I went to see my bank manager to get his advice. 'We don't buy and sell shares,' I was told. 'You'll have to go to a stockbroker.' In any case, I wouldn't be buying Woodside shares; there are far safer options available for investing your money. Also, Woodside shares are not 5c anymore; now they're $8.90. I was a babe in the woods, wasn't I.

I left his office confused, disappointed, ashamed of myself. How could I have missed such an opportunity? If I'd asked Mike or any of the office staff, they would have told me to go to a broker, and probably referred me to one. What an idiot! As it turned out, I could still have bought the shares, and done well with them because they kept going up, and some years later got to just shy of $100. I was learning, but it was costing me. In later years, we had the computer, and an online broker account, online

Chapter 2 - Australia. The Early Years

banking, money transfers, and so on. How the world has changed in 50 years!

Back in the desert, the Pexa well was nearly half-way through when there was a breakdown. This turned out to be a major breakdown, and parts were going to have to be trucked in from Roma. That meant the rig was going to be closed down for two to three weeks. What were Ned and I going to do? We were going home, that's what we were doing, and coming back when they were ready for us.

Ned had been on the radio to Charlie Jones, and Charlie wanted to know what was going on. He invited us over to his place in Adelaide. He would pick us up from the plane. Charlie was tall, but not heavily built. Silver grey, crew-cut hair, and piercing blue eyes, and a sharpness about him that said he'd lived by his wits for his whole life. We got to his place, and settled in to have a few beers, and talked about Richter Bawden and their problems. His wife appeared; it was obvious that she didn't approve of his drinking, and we certainly weren't staying for dinner, so the meeting was cut short, and we headed back to our hotel, The Haven Inn, at West Beach.

I had a couple of weeks to myself, and I spent it going around looking at yachts, and marinas. I knew fairly well what I wanted, and how much I was prepared to pay, but there were other things to consider, like where I was going to keep it; and whether I would be able to live on board — something I badly wanted to do. A few days down among the yachts and the marinas was like a holiday after weeks on a roaring drilling rig set in the desert.

The other thing I did was go to the races. I was away for a lot of the time these days, and this was my first time for a few weeks. It was Randwick this Saturday, and I got out there bright and early and had a couple of bets on the early races. Not much doing. I was hanging around between races, doing not very much, when something caught the corner of my eye. There was a bloke walking my way, and as he neared, I realised it was Ronnie Hamilton. He hadn't seen me, and it looked like he was going to walk straight past me. 'G'day, Ron,' I said. 'How're ya doin'?

He stopped and did a bit of a double take. 'George,' he said. 'It's you. I wondered if I'd ever see you here. What's happening. 'I wasn't travelling too well today, I told him. 'I need one of Bob Backlash's tips.' He laughed. 'We're not in Cloncurry now,' he said. 'C'mon, Ron,' I said.

'Waddaya know? Give me something, just for old time's sakes', he said.

'I'll give you #5 in the last… for Chrissake, don't tell anybody I told you.

'No problem,' I said.

As he walked away, he wished me good luck. I got out my form guide and had a look at the last race. There it was #5, 'Band of Hope': third race in from a spell; did nothing in the first two. Won at 10 furlongs last year. Looks like it could have been set up for this race today; pre-race betting 12-1.

If it's any good, it won't start at 12-1. Will I have a plunge on it? I was losing on the day, and if I could snag this one it would set me right. The second last race was over, and the bookies were paying out winning bets. Then the boards cleared, and the betting went up for the last race. Band of Hope 12-1. Immediately there was money coming from the members side of the rails, and the boards went down to 10-1. Time to move. I tried to claim Arthur Singh for $100 each way, but he would only give me 8-1, all the bookies were dropping the price, 7-1, 6-1. I had a look over at the TAB board. It was still showing 10-1 so I went over and had $100 each way.

There was a lot of money went on that horse and it started at 5-1. The jockey had it well-placed throughout the race and as they straightened up for the last 400m he was well positioned. At the 200m he was in front and going away. The favourite came at him after that but wasn't good enough, and he won by a length. Yippeee!!! I'd won nearly $2,000 on the race, and over $1500 on the day.

I queued up at the bookies stand to collect my winnings and I just had the feeling that someone was watching me. I turned

Chapter 2 - Australia. The Early Years

around and looked at the queue for the next-door bookie. I saw a face turning away as I looked. Who was that? To put it all in perspective, I went to the races again the next week at Rosehill and tried to be smart. There were no tips this time, and I finished up with one win, two placings, and two losses, which left me just in front for the day.

The next Monday, we got the news that Richter Bawden was ready for us, and we had to get down there, quick smart. Ten days and the well was finished. Nothing interesting; just a couple of minor shows. It was announced that Pexa would not be doing any more wells. They were cancelling the last one. I got a message from Mike to be sure and bring all the gear with me as he had another well lined up for me. The general feeling was that exploration was coming to an end in Australia, they'd found everything that was going to be found for the time being. The North Sea was about to start up, and that was going to be the focus. Ned was talking about the North Sea. He'd been told he was going up there, and he asked me if I'd thought about what I was going to do when it all went quiet. I hadn't really thought about it very much. Mike had told me he had more work for me, and I'd have to see where that went before thinking about anything else. Also, if I went up to the North Sea for several years, that would be the end of my dreams of owning a yacht and sailing. The North Sea is cold even in summer. Anyway, Ned gave me his number in Brisbane, and told me that, if I wanted to apply, he'd put in a good word for me. I thanked him for that, and that's where we left it.

When I got back to Sydney the office was abuzz. There were two wells to be drilled in Queensland, one of them in the Rolleston area, and the other down the Thomson River, south of Longreach. Mike Gahan, and one of the geologists, whose name also was Mike, had some skin in both of them, so there was a bit of extra stress around. We were taking the cars this time, driving up, and were due to leave in a few days. Mike, the geologist, had already gone up there in his private car and I was to follow in a company car with all the gear. Richter Bawden were doing the drilling, and the well proceeded normally. No oil was found, and no gas, but the drill did pass through several thick seams of coal. Top class coking coal, and this area became a massive open cut coal mine in later years, so maybe something came out of the well after all.

When the well finished, a message came through for us to move over to Longreach. The drill rig would be moved over to the new site down the Thomson River fairly quickly, and it wasn't worth going all the way back to Sydney, just to come all the way back in a few days' time. Mike Gahan, my boss, wanted to be there when the well finished and I had to drive up to Longreach airport to pick him up, and there wasn't much to report on the way back to the rig. There wasn't much else to do except pack up all the gear and head back to Sydney. Come into the office next week, was the message. I've got something for you to do.

The something was a survey. A landowner up in the Gloucester area in NSW had contacted Mike with a request to survey some of his land in the hope that there might be some mineralisation there that might be of value. There were going to be three of us plus gear, and we would be taking two cars. We left on a Monday morning and drove up to

Gloucester and checked ourselves into the Gloucester Hotel. On Tuesday morning, we went to see the landowner who took us around and showed us the land to be surveyed. The land was quite steep and rough, and we had to establish a grid to reference all the samples we picked up. It was Wednesday afternoon before we finished the grid and started picking up the soil samples.

The next morning, we had an early breakfast and got going. It was a tough day, the geologist wanted to get it finished that day, but the ground was difficult, and the right samples hard to get. By 5pm it was starting to get dark. We gave it away for the day knowing we'd have to come back tomorrow for a couple of hours to finish it off.

After the evening meal, we repaired to the bar for a few quiet ones. Get to bed early, was the message; start early in the morning.

Chapter 2 - Australia. The Early Years

My workmates had disappeared; surely, they hadn't gone to bed yet? I heard voices as I went down the corridor, and there they were in the coffee shop, and Donna was serving them coffee. Evening all, I said as I walked in. A cup of coffee is just what I need right now.' It was just right, and what followed was an intense night. Suddenly, the dawn was creeping through the windows, and it was time to go. I crept down the corridor hoping to creep back into bed before the others got up. Too late, they were already getting ready for breakfast. 'Where've you been? 'Just out for my morning run… 'Ha, ha, a likely story. We're going down for breakfast, are you coming.

A quick freshen up, and I was there. Did you hear all that noise last night? Round about 11.30pm. sounded like someone run amuck with a hammer. I'd not heard a thing. After breakfast, we're getting ready to leave, and the boss says, make sure you take everything with you. We won't be coming back here tonight.

Walking over towards the cars, something doesn't look right, and as we get closer it becomes obvious. The doors have been kicked in. 'That was the banging that you heard last night.' The boss was furious. 'That was the guys you were drinking with last night,' looking at me as if I was to blame. Yeah, I hear what you say, but you're never going to prove it, unless some witness turns up. I went round and checked all the doors. They opened and closed properly and locked. They were secure, and the cars could be driven.

'We're going to have to report it to the police,' I said. 'It's part of the insurance process.'

He wasn't happy about that, either, but we went to the police station, and did the right thing. The constable got his eye on me. 'You're not driving, are you?' he said.

'No way. Not today. 'The rest of the work was completed, and we got back to Sydney by late afternoon. I was to take the samples with me and deliver them to John in Kent St on Monday morning. It was going to be a quiet weekend for me.

John was happy to see me on Monday morning. Stay a while and have a chat he said. You're not going to the movies, are you?

No, not today, I said. I'm thinking of going down the office to see what's happening. 'Don't go down there now,' he said. 'I've been on the phone to Anne (the secretary) this morning, and its chaos down there. Looks like it's closing down. The oil exploration industry has closed down in Australia, and there is none of that kind of work anymore.

That geologist has been badmouthing you to Mike,' he said. 'The fieldhands, which means you, are being made redundant. You'll have to go and get yourself a cab driver's licence.' This was something John and I had talked about before. He had a brother who was a cabbie and was earning a motza at the moment. I tried to imagine myself as a cabbie. I hated cab drivers, hanging around at all hours preying on drunk people, overcharging, going the wrong way. Never in my wildest dreams had I ever considered something like that. Give Anne a ring now John was saying and see what she says.

Anne was sweet—she was always sweet— 'Sorry to have to tell you, George, but we're closing down here, and you're being made redundant.

'Mike wants to see you tomorrow morning, and we'll fix up your pay, etc. Could you come in tomorrow morning?

No problem, I said. Look forward to seeing you. It didn't take long, Mike was very conciliatory, the industry was closing down in Australia; there was no work for me anymore. The cars were being repaired, no problem. He wasn't blaming me for the damage. He wished me well.

Anne had all the paperwork organised. The office was empty; they'd all gone, except Anne. She was going at the end of the week. I went down to Kent St. I couldn't go without seeing John one last time. He had some work to finish off then he was finished too. He wished me well. Go and get that cab licence, were his last words.

Chapter 3
Dorado

Getting a cabbie's licence wasn't difficult. I went down to the Transport Registry at Roseberry to make enquiries. They were happy to see me; they needed cab drivers; there was a shortage at the time. I signed up for a two-hour lecture, at the end of which we were given a pile of stuff to learn about how to get around Sydney. There would be an exam based on all this information, and we had to get 80 per cent-plus to pass. I knew hardly anything about Sydney, but I managed to pass the exam by swotting up on the given information, and lo and behold, there was my tentative licence. My first shift started from a base in Bondi, and my first fare was from Bondi Beach to Central Railway.

The first few weeks were a steep learning curve. I'd been told by any number of people that the way to learn was to get out on the road and work, and that was the way it was. As time goes by, the confidence builds up as you learn how to drive and deal with people. Little did I know then, but the cab game was a thread that was going to run through my life for the next 50 years. It was a one-man operation, and it was up to you to work hard and make the most of it.

I soon discovered that it suited me fine to be free of offices and bosses. Lots of people think they'd like to try the cab game, but very few last more than a few shifts. It's a cut-throat, dog-eat-dog industry, where your workmates are your competitors, and you are very much on your own to deal with it all. Dealing with the public can also be very difficult, and that one-in-a-hundred can ruin your day, or worse. With experience, you can often pick that problem passenger before you pick them up, and just drive past them. If you do get a bad one, you're stuck with them. Just get them out of the cab the best way that you can and get on with your day. Don't let them upset you.

I was settling in. A few weeks had gone by, and I was feeling more confident. It was a good time to be driving; the Viet Nam war was still going, and American soldiers were coming to Sydney on their R&R leave, jumbo jets full of them every day. Sydney nightspots were full of them every night. The gals were there to meet them, of course, and the Americans were such gentlemen... they felt that they just had to see the girl home safely. So, it was taxi out to the far suburbs or even beyond, wait for a few minutes, then back to the city. How long could that go on?

Not very long, as it happened. I started off driving seven days a week, but that didn't last long. It didn't leave any time for other things, even basic things like doing the washing, and having some kind of social life. So, I cut back to six shifts per week, and that was better. I could keep that going for the time being, anyway. There were two shifts in the cab game—3am-3pm, the so-called day shift, and 3pm-3am, which was the night shift. I was a night shift driver for most of my career simply because that's where the money was. It is also true to say that it is the night shift where most of the hassles were, but for me it was the money, especially in the early years.

I wasn't living at Armstrong St then. I was given a notice to vacate as it was being redeveloped. I'd discovered Balmain and Birchgrove by being a cab driver, and I fancied living there. I had no trouble finding a room in Louisa Rd in Birchgrove. People today might be surprised to learn that in those days Birchgrove/Balmain was a cheap place to live, and Louisa Rd had several boarding houses. I was earning lots of money, and my bank was building fast. I now had more than enough money to buy that yacht that I'd been dreaming of for years. Was It really going to happen? I started looking in the 'Herald' and buying sailing magazines which advertised yachts for sail. I knew where most of the yacht clubs and boatsheds were through my work as a taxi driver. One day I saw an ad, placed by the Castlecrag

Chapter 3 - Dorado

Boatshed on Sydney's north shore: 35'steel yacht, cutter-rigged sloop, raced in 4 Sydney/Hobart races, finished 69th out of 96 last year'. That sounded very much like the kind of thing that I was after. I headed down to Castlecrag to have a look. Down that steep narrow twisty driveway, with the boatshed at the bottom, and there was 'Dorado', tied up alongside a pontoon, and open for inspection.

It was love at first sight. This was the yacht I had dreamt of, and here it was, and I had to have it. I could have asked questions about design, and sailing performance and other things, but I was new to yachts and sailing, and almost completely ignorant about such things. The fact is that was I was lucky. I got a good, solid cruising yacht, almost ready to go, plenty of sails, all in good shape. I still had some of these sails in December '83, when the disaster happened.

Part of the sale deal was that the previous owner would come down and do a day sail with me, and I was very keen for this to happen. John Lake was an auditor, and lived in Tasmania, however, he travelled to Sydney regularly on business, and he was able to organise a Saturday morning to come up to Sydney. I picked him up at the airport, and we went straight to Castlecrag. This day was invaluable to me. I got my initiation to sailing. How to hoist the sails, trim the sails, recognise the rule of the road, start the engine, etc, etc, etc. For the first time, I got the feel of a boat moving through the water under sail. That special thrill was something I never lost in 25 years of boats and sailing. John's visit got me started; the rest was up to me.

The next move was to move the boat closer to where I lived.There were plenty of boatsheds and two marinas in Balmain, and after a few knockbacks I was lucky enough to get a marina berth at the Balmain Marina, run by Alf and Clare Taylor. Alf and Clare became friends and were very good to me in the years to come. The immediate problem was to get 'Dorado' from Castlecrag to Balmain: the first voyage; the first challenge. George Howie was working in Sydney at that time and wanted to come with me if we could make it Saturday morning. No problem, I needed all the help I could get. Sylvia kindly dropped us off at Castlecrag, and the boatshed ran us out to the boat, which was on a mooring. Follow the procedure and get the engine started… Water coming out of the exhaust? Good. Drop the mooring, engage the gear, and off we go.

We motor down to the Spit Bridge. We've missed an opening and have to wait for another hour. In the meantime, I get the mainsail out and 'bend' it to the boom, fit all the slides to the mast, and attach the halyard. The mainsail is ready to hoist. Past the bridge, down Middle Harbour, approaching Middle Head. I decide to hoist the main, but I'd forgotten something. When not in use, the main boom is attached to the backstay with a strop and a shackle; it stops the boom moving around when there is any kind of sea running; and I'd forgotten to release it. As a result, the sail filled immediately when I hoisted it in the fresh breeze, and the boat dived up into the wind almost uncontrollably.

We fought with it for 10 minutes or so till it was obvious that we couldn't go on like this. The backstay was taking a lot of unnecessary punishment, and this was Saturday morning. There were fleets of racing yachts around and they had right of way over us. There were ferries buzzing backwards and forwards, which also had right of way. So, I dropped the mainsail, and we motored on till we rounded Bradleys Head, where the breeze eased a bit, and our course was almost due west. I rehoist the mainsail, properly this time, and set a working jib, and we had a comfortable sail up to Balmain. No harm was done but we'd given ourselves a bit of a stressful day.

The next few months were a time where I learned how to sail, and to handle the boat. George and I went out on the harbour nearly every Saturday and Sunday and learnt how to handle the boat in various conditions. As we grew more confident, we started to go out through the Heads and encounter offshore conditions. The passage through the heads was always the roughest part, with its cross seas and tidal currents. I sometimes got the 'mal de mer', which amused some of my friends, who looked at me askance saying, '…and you want to sail round the

Chapter 3 - Dorado

world?' Eventually it stopped and never bothered me again.

Sydney harbour was a large expanse of water, or so it seemed to me in those days. There were almost no dangers, and the tidal currents were quite modest compared to some parts of the world, and so it was quite safe to sail almost anywhere. It wasn't long before we got the hang of it and started looking at offshore trips to get out of the weekend traffic. Pittwater was first. A good day sail up the coast. Anchor round behind Barrenjoey, and spend the night, then sail back to Sydney Harbour the next day. Terrigal on the central coast was next. The anchorage was a bit iffy and very rolly. It was a very uncomfortable night (mal de mer territory for some people). On the way back, we encountered a sou'-east change, and had to go into Pittwater and hide in Refuge Bay. The next day, the wind had gone back to the nor'-east, and we fairly raced down the coast and into Sydney Harbour.

Our next voyage was a little more ambitious. Port Stephens is about 100 nautical miles up the coast from Sydney, which meant that it was going to take us around 24 hours to get up there, and maybe about the same to come back. We were going to have to pay a lot more attention to navigation, and have all the charts for the coast, and for Port Stephens itself. We set off from Sydney on a day when the nor' east sea breeze had set in, and we tacked up the coast all day and into the night. We had to contend with the East Coast current, which runs to the south 12 months of the year.

Around midnight, the breeze dropped away completely, and we started the auxiliary motor to keep going. There were lots of fishing boats around, and we had to keep a close watch for them. We were off Newcastle when a light westerly breeze began to come off the land. This is a sure sign of fine weather, and a couple of hours later the westerly faded out and the sea breeze started up. We made it around into Shoal Bay by early afternoon and anchored just off the beach, and close to the Country Club Hotel.

A southerly change came through overnight, but we were

quite snug in our anchorage. In the morning, the weather had improved, and the wind had eased off to the sou'-east. We waited till about lunchtime before setting off. The sailing conditions were quite good with a sou'-east breeze of 20 knots, and with the help of the current this time we were back in Sydney early next morning.

Just to balance things out, we decided to take on a voyage to the south, down to Jervis Bay which is about 80 nm down the coast. It is a large bay, and there is a Royal Australian Navy base there. The base is off limits to civilians, of course, but there is other settlement there, and the village of Huskisson, which is set on the entrance to a small creek. It offers safe anchorage in the creek, and the usual services. The voyage south was quite fast with a sou'-east breeze backing to east, then nor'east quite fresh, and we passed through the heads into Jervis Bay the next morning.

We stayed around Jervis Bay for about 36 hours. The northerlies grew quite strong, and we delayed our departure for a few hours to see whether the southerly would arrive. In the end, the northerlies eased, and we decided to head out knowing that the southerly would arrive soon. We'd barely got clear of the heads when the southerly arrived with a bang. We had to reduce sail. We reefed the main, and set a working jib forward, which kept us fairly busy for a while. Things settled down a bit after that, and we started to make good progress. It was rough, the roughest I'd seen so far. The wind was gusting up to 35 knots-plus, and the sea was really building up behind us. After a few hours the wind began to ease to the sou'-east, and by the time we came around South Head and entered Sydney Harbour it was only a light breeze.

One other trip I did about that time happened in the week between Christmas and New Year. Everybody was away on holiday. Even the racing boats were away in Hobart. I could have the harbour to myself and practise a bit of single-handed sailing. The weather was stable, and there was no threat of any

Chapter 3 - Dorado

southerlies. Down the harbour. The nor'easter was in and the weather was set fine. Rounded Bradleys Head, and the breeze was right in our teeth, so it was tack… tack… tack all the way to the Heads.

I was enjoying this, and there was plenty of time, so it was an easy decision to make. Keep going, out through the Heads. I wanted to experiment. Could the boat be set up to sail by itself for a period of time. Some boats could, by getting the settings just right. Many of the roundthe-world yachts had self-steering gear, and some of them were very expensive. I already knew that 'Dorado' could have a lot of weather helm at times. When sailing to windward or reaching, she would heel over, and once she passed a certain angle of heel the weather helm would increase rapidly, and she became very hard to control. If I took a reef in the main that would help it a lot, and that was to become standard practice.

I'd been so engrossed that I hadn't been paying attention, and that was going to cost me. I'd noticed a small cloud away down on the southern horizon. No problem I thought, there's nothing in the forecast, and I carried on. Next time I noticed, it was much closer, and seemed to stretch right across the sky. This is looking threatening, I thought. The nor'-east breeze held; the cloud neared. I could see through behind it, and the sky was blue. I started to relax. This wasn't going to be too bad; it had nearly passed us. Wham! No warning. A powerful gust of wind from the south. The boat went over, mast right in the water. I nearly went too. I just managed to grab the starboard winch and held on with grim determination. The gust started to ease, and 'Dorado' started to right herself. I scrambled up as the wind came back, a steady 40 knotsplus. The sails had to come down, no matter what, or we were going to lose the mast. It took some time to get everything under control, and I then retreated below to take stock.

I had just experienced a 'line squall'. I'd heard about line squalls, and read about them, and I should have been warned by that approaching cloud. I'd had a very sharp lesson, and I wasn't out of it yet. The wind was still gale force, and the seas were building up and starting to break. Spray was flying everywhere. I figured we were about 10-12 nautical miles out to sea, and it was now dark, and I could see the Macquarie light.

I thought the current might be pushing us south, away from Sydney Heads, so I started up the auxiliary. Progress was slow, but at least we were headed in the right direction. Later on, the breeze eased to 25-30 knots, and I reset the jib. Progress was much better then, and we passed through the heads just as dawn was breaking. Inside the heads the breeze quickly dropped away to light.

That's the thing about line squalls: they happen out to sea and rarely affect the land and are very hard to forecast. I reflected on the night as we motored down the harbour. I'd had a sharp lesson. I should have been ready for the line squall. It wasn't as if I hadn't had any warning. I nearly went overboard and was lucky there wasn't any serious damage to the boat and gear. The sea is a hard taskmaster, and you don't get too many second chances.

I was learning fast. Our next voyage was going to be to Lord Howe Island. George Howie and I had been talking about it for some time. Lord Howe was about 480nm nor'-east of Sydney and would require some accurate navigation. With Alf's help and advice, I'd got myself a good quality sextant and Alf showed me how to set it up and check its accuracy every time before use. I got a copy of the latest Nautical Almanac which contains information about the movements of the sun, moon, and the planets for the current year. Time was going to be measured by my Citizen watch, which was very accurate, and could be checked against the atomic clock.

The atomic clock broadcasts on several frequencies (5, 10, 15, 20, 25mhz). Any half decent radio receiver will receive it on one of these. I have never failed to receive their signal anywhere in the world. I'd done all the mathematics involved in sun sights at school, and so I understood what it was all about. All the hard bits had already been worked out in the Nautical Almanac, however, and the job for the navigator was to get an accurate

Chapter 3 - Dorado

sight.

That requires practice and practice and practice. Standing on the heaving deck of a yacht, you might see a true horizon for three or four seconds every 30 or 40 seconds, and you have to bring the sun down to touch the horizon in that brief time. Practice makes perfect. I might point out that this is the early 1970s, and GPS did not exist, and even the forerunner to GPS called Satnav was not available for small boats.

Alf also gave me a complete set of books with precession tables for all navigational stars up to the year 2002. I found them very useful, indeed, during the circumnavigation that was to come. Other navigational aids include the log, which is the equivalent of the speedo on your car. It's not a simple 'diary', as many might expect. It's a device that's bolted on the hull of the boat, with a readout dial in the cockpit and it tells you how fast you are going, and how far you have travelled since you last checked. Very important for dead reckoning when it is cloudy, and sights are not possible. Also, the hand bearing compass, useful for taking bearings on coastal features (important in dead reckoning exercises), and as a standby compass in an emergency.

If you are navigating by dead reckoning, something you must take into account is leeway. As a sailing boat moves through the water, it makes forward progress. It also makes sideways movement, which is known as leeway. Leeway can vary quite a lot depending on sea conditions, weather conditions, ocean currents, etc. You can never know precisely how much leeway your boat is making at any one time, but with experience you can estimate it. It could be important in a tight situation.

Alfie Taylor ran the Balmain marina with his wife, Clare. He had had an interesting life. His family were Balmain people, and his father had worked around the wharfs all his life. Young Alf was looking for adventure and ran away to sea on the mission ships that traded around the Pacific Islands. These ships were sailing ships, albeit with a powerful auxiliary. He eventually became the mate on one of these ships. In the 1930s, he made his way to the UK and joined the merchant marine. During the Spanish Civil War, he volunteered to crew a ship that was delivering armaments to Barcelona. When they got to Barcelona it was under attack and they couldn't enter the harbour. They were hanging around trying to decide what to do when a warplane came over and dropped a bomb on them. It was a king hit, and the ship blew up. Alf was on the bridge at the time and the blast blew him into the water 100 metres away. He swam ashore and was promptly arrested and thrown in gaol. The battle for Barcelona was fluid, and just when it looked like one side was going to win, the other side came back and retook the city. Alf was released from gaol and managed to board a British destroyer which was looking for survivors. Alf stayed in the UK and during the 2nd World War he crewed on the Atlantic convoys. He was sunk twice and survived to tell the tale.

As if this wasn't enough, Alf volunteered for another dangerous job. The British were getting another convoy together to send armaments to the Russian port of Murmansk up in the Arctic circle. Of course, the Germans were waiting for them, and Alf was sunk again. He was rescued after about 20 minutes in that icy water and taken to Wick, in the north of Scotland, where he was hospitalised for several weeks while they got the oil out of his lungs and stomach.

Alf was repatriated to Australia after that episode. He'd had enough of the North Atlantic. By this time, war was raging in the pacific, and Alf was roped into crewing ships running supplies into the war zones. He and was sunk twice by the Japanese. After the second sinking, the survivors were on a life raft for 21 days before being rescued by the Americans. (One of Alf's fellow survivors from that life raft was a guy called Norm Simmons, and it just so happened that Norm lived in Balmain. It was fun and games when Norm came a' calling.)

By the end of the war, Alf had his master's Certificate, and he continued his maritime career as a captain on ships trading on the Australian east coast. Alf knew people in high places, and in 1950 he was offered a business opportunity to bring sandblasting

Chapter 3 - Dorado

for ships into Australia. It went well, and Alf made his fortune in the next 10 years. He then sold the business for a huge profit and could have retired. Instead, he bought an old hotel in Balmain and turned it into a sailing school.

The sailing school was residential, and young people involved in the merchant marine industry could go there and study, and sit exams for their various marine tickets, right up to masters. During that time, he had a star pupil whose story he often related when he'd had a few scotches. He was German, and Alf always referred to him as the 'Kraut'.

He had come to the Pacific as a boy with his parents before the First World War when the northeast part of New Guinea was a German colony and Tonga had quite close ties with Germany. Like Alf, he had run away to sea on the mission ships when he was young, and eventually crewed on the clipper ships which carried the wheat and wool from Australia to Europe right up to World War II. He'd never been master of a clipper ship, but he'd always wanted to be one, and he came to Alf to study and pass the exam, which he did. Alf reckons it was the last one ever issued.

I'm telling you all this because I actually met this guy, years later in Tauranga in New Zealand. I was berthed at the old yacht club up the river, just before the railway bridge. One day I see an elderly gentleman (he must have been well in his 80s) coming towards me and staring right at me with these incredible eyes. I was busy doing something else and I didn't want to talk, but he persevered, and I suddenly realised that this was the 'Kraut', Alf's star pupil. We spoke for ages.

He came back two more times and spoke of the clipper ships and Cape Horn. I'd been very cagey about my plans for Cape Horn, and I hadn't told anyone, but I think he knew what I was planning, and wanted to pass on the knowledge that he'd gained from many years' experience in the Southern Ocean. I enjoyed his company and wrote a letter to Alf and Clare saying that I'd met him.

In the 1960s, Alf sold the sailing school, and bought The

Balmain Marina, which was supposed to be something to do in his retirement. Alf wasn't made that way, of course, and it was soon a full-time job for him. Alf had met Clare at the end of the war. They were married, and there were no kids, but she was kept busy running Alf's businesses. Alf could be grumpy at times, but Clare was just a lovely person, and I never had a problem with her.

There were actually two marinas in that part of Balmain, and they were next door to each other. The second was Cameron's Marina, and there were quite a few yachts there that allowed people to live on board, and so we were a little community. One guy in particular was Frank McAlister. He was a Liverpool lad, and he had built his own 40ft ketch. We became good mates and drinking buddies.

If you own a yacht and if you're going to sail it regularly, you have to keep it in good shape. In particular, the hull below the waterline needs regular attention. Weed and barnacles will soon become a problem and will badly affect sailing performance if you don't do something about it. The answer was to slip the boat regularly, clean the hull and apply new antifouling. In common with most boatsheds and marinas, Alf had a slipway available for hire, and I slipped 'Dorado' there at least twice a year. Sydney has a modest tidal range (2.0m high water springs) so these things have to be booked well in advance, and you have only a short time to get the job done. It is hard work, of course, but it is a labour of love. I have to take time off work for this, but that's OK; the boat always comes first.)

George Howie and I had decided to go to Lord Howe Island in February. We reckoned it would take us about five days to get there, rest up for a few days, then five days back. Three weeks would cover it nicely. It was going to be my first test of offshore navigation, where I had to rely on getting my sights, and getting them right.

The day comes, and we're ready, everything is checked and

Chapter 3 - Dorado

rechecked. Down the harbour, through the heads, and out to sea. The breeze is light/moderate sou'-east. We've got the mainsail and genoa set, and close-hauled. We're just about laying the course for Lord Howe. I'm getting sights: morning sight, noon sight for latitude, and afternoon sight. In the meantime, we're reading the log at regular intervals, and recording our DR (dead reckoning). We're doing three-hour shifts on the tiller; all hands on deck for sail changes.

Overnight, the breeze backs a little to east and freshens a little. At dawn, we take down the genoa, and replace it with #2, leaving the main as is.

Morning sight, then noon latitude, show we've made good progress.

Afternoon shot confirms we've covered around 200nm from Sydney, c. 300nm to go. Breeze has freshened further in the late afternoon, so we put a reef in the main, and change to #3 jib. We hope that will get us through the night.

In the morning, things are not looking so good. The breeze has freshened to 25-30 knots and has backed to the nor'-east. The sea is getting up, and we're hammering into it. Spray is flying everywhere, and navigational sights are very difficult. I did get some questionable sights in the end... progress not so good, especially in the last 12 hours. Now we're heading just about due north, but if we went on the other tack, we'd be headed due east.

The situation was this: There was a trough of low pressure in the Coral Sea when we left Sydney. That's quite common in February, and very often they come to nothing. Sometimes, they intensify and become tropical cyclones, and that is when they start to move. Sometimes, they head for the Queensland coast, where they do untold damage and flooding. Sometimes they head sou'-east down the Coral Sea, and into the Tasman Sea passing east of Lord Howe Island, and it seemed that that was what was happening on this occasion.

The ocean forecasts were full of gale warnings, and even storm warnings for the Tasman Sea for the next few days, and the nor'-east wind that was currently hammering us was going to increase to full gale force in the next 24 hours. For us to carry on was just insane, and the only sensible thing was to turn back and make a fair breeze of it. We were actually about a day's sail to Port Stephens, and that's where we headed for shelter. I got in some practice taking navigational sights in rough weather, and we arrived in Shoal Bay late the next day and anchored near the Country Club Hotel. The decision now was to wait until the cyclone passed to the southeast. That would take more time than George had available, and we decided to abort Lord Howe for now, but to rearrange the trip for later in the year.

The trip back to Sydney was interesting. We left Port Stephens with a nor'-east breeze force 4-5 and were flying along with help from the current. After a couple of hours, the breeze backed to east force 5-6, and not long after that it went to sou'-east force 6. It was getting fairly rough, but we were holding on with a reef in the mainsail and #3 jib. Then things started to really deteriorate. The breeze now backed to south force 7 and was now forcing us down onto the coast. We could see the Norah Head light, and a quick bearing on that told us what we needed to know. We had to tack straight away, and we did, and not a moment too soon. As we settled down on the starboard tack, we could hear the roar of surf pounding on the coast behind us.

We held onto the starboard tack for about four hours by which time we were safely out to sea. It was tough going, and when we changed to the port tack it crossed our minds that we could make it into Pittwater on this tack and rest up for a while. However, the breeze veered to the sou'east, and eased off to force 5, and that allowed us to lay Sydney Heads, and get back into Sydney Harbour. I've noticed over the years of living in Sydney that the cyclones will never come anywhere near Sydney, but when they are around up north, we very often get unsettled unstable weather with strong to gale force SE-E-NE winds offshore, and very humid, and sometimes rainy weather on the coast.

It was early November when we made our next attempt on Lord Howe Island. There was to be no turning back this time.

Chapter 3 - Dorado

There had been a few sceptics the last time, and we needed to shut them up. We left Sydney Harbour on the tail end of a southerly and made fast progress for the first 24 hours. The breeze then backed to about sou'-sou'-east, and a few rain squalls came through. The breeze would increase to 25-30 knots during the squalls and force us to reduce sail, only to increase it when the squall had passed. This was giving the deckhand plenty of work to do. The deckhand was also the navigator, and I was quite busy between squalls taking sights and keeping the dead reckoning up to date.

As night approached, the squalls decreased, the breeze backed to sou'east 15-20 knots, and we were sailing comfortably at six knots on a broad reach. We were doing three hours on-three hours off on the tiller. I was the navigator, and the deckhand, and George was the selfappointed cook. The breeze slowly eased overnight, and by dawn it was down to five knots, and progress was slow. We had covered 350nm in the 48 hours, and still had approx. 150nm to go. There was a period of calm then a light breeze picked up from the north. It quickly freshened up and backed nor'-west, and we were soon flying along at seven knots. By noon the breeze had backed to west and was gusting up to 30 knots, and we were under much reduced sail.

The sea conditions were now fairly rough, but the sky was clear, and I managed to get some decent sights, which put us a bit more than 100nm from Lord Howe. In the late afternoon, the breeze backed away to the southwest and started to ease. We made slow progress overnight, and the breeze faded away completely, but as the dawn came up, what a pleasant surprise: the twin peaks of Mt Gower and Mt Lidgbird, and Balls Pyramid to the south. 'Hey, look at this,' I yelled out. 'We're nearly there!' Mt Gower is nearly 3000 feet (c. 900m) high, and Mt Lidgbird is about 2500 feet (760m), and so are visible, while the rest of the island is just starting to appear over the horizon. We were still about 25nm away, and still had several hours of motoring to get there.

Lord Howe Island has a coral reef on its western side, and

that creates a large lagoon, offering safe anchorage in all weather. We had rung the harbourmaster before we left Sydney, and he had advised us to go to the north end of the reef where there is a passage through that leads to a safe anchorage. He actually came out to meet us and guided us through to a large hole with a mooring in the middle.

Lord Howe is just about my favourite place on planet earth. Remote and isolated, hard to get to, and far away from the hustle and bustle of everyday life. Beautiful, and with a great climate, it is just the place to come to, to relax and get away from it all.

This was 1973, and the Sunderland flying boat was the only way people could get to and from the island. It landed and took off from the lagoon, and that meant that it was highly dependent on wind and tide. Cancellations were common. A small supply ship came once a month but carried no passengers. In the late 1970s, an airstrip was built, and Qantas commenced daily services. It is a much better service than the flying boat but can still be cancelled due to weather conditions. The number of people on the island is strictly controlled. It is virtually impossible to become a resident there, and visitors are strictly controlled; no more than 400 visitors at any one time, and they must have arranged accommodation. There are no pubs or clubs on the island, but the tourist hotels have limited supplies, and George Howie and I discovered that there was an RSL club open four hours per day.

We never intended staying more than a few days on the island, however a south-west gale set in, and we delayed for a few days. Finally, it eased and backed to the south, and that was our cue to get going. We had variable conditions on the way back, including a westerly gale with very big, confused seas. We hove to for 12 hours during the worst of that blow, then carried on after the wind eased and backed to the south-west. It was a slow trip compared to the trip out, and it was nearly six days before we arrived back in Balmain. It was all part of a learning process, of course. Learning about the boat and its gear, life at sea, navigation, and discovering Lord Howe Island.

Chapter 3 - Dorado

When I brought the boat to Balmain and began living aboard, I was the only one on the Balmain Marina. There was a houseboat in another berth, but there never seemed to be anyone there. In fact, it was up for sale, and eventually somebody bought it, and moved on board. Her name was Kay Brooks, and she was a Qantas international air hostess.

She had lots of friends, and when she was there, she had people visiting her regularly. Clare warned me about her. She's a hussy, Clare told me one day. Watch out for that one! Kay was English, and as we were neighbours, we got to know each other. She had a boyfriend in London, she told me, and she was trying to entice him to come to Sydney. His name was Steve, and he was a bodybuilder, she told me. He takes part in competitions such as Mr World, etc. 'My hostie girlfriends go and visit him when they're in London,' she said. 'It's not ideal, is it. I'd prefer it if he was here in Sydney'.

I hear what you say. She'd go away on her working trips, usually 2-3 weeks at a time, and when she came back it would be with one of her male workmates. She wasn't missing out on anything, was she. Qantas used to fly to New York and the Bahamas in those days. Flight crews had glamorous lives, that was for sure. One day, she was there on her own, and we started talking. 'Do you know where Refuge Bay is?' she asked me.

'Yes, it's up in the Hawkesbury,' I said.

'Could you take me up there?'

'Have you ever been sailing before?' I asked her.

She said she had, 'and I don't get seasick'.

It's a day sail from here, I said, always depending on the weather. She said she had friends going up there the following weekend.

'They've got a big boat, and they're having a barbecue,' she said. 'I missed out on an invitation, but I'd like to go up and surprise them.'

I'm not the best in these social situations, but I couldn't

refuse her, could I! It wasn't the sailing that was going to be the big challenge for me this time.

The sail up was quite fast, we found a light sou'-east breeze down the harbour, which took us out of the heads, where it backed up to the east and started to freshen. In the late afternoon, it got quite fresh from the nor'-east, and we had to tack four times to get around Barrenjoey Head. The rest was easy, straight up Broken Bay and into the Hawkesbury to Refuge Bay. We found her friends without any trouble and anchored nearby. The party was already underway, and I fitted in no problem. I was George, the Sydney taxi driver. There were quite a few gays or bi's there, but that didn't seem to be a problem. This was the 1970s, and there wasn't the polarisation that started up after 'aids' arrived in the 80s.

The party went on into Sunday, and we stayed with it, but I knew that there was a southerly change due on Monday afternoon, and I wanted to be back in Sydney Harbour before it arrived. I wanted to leave Refuge Bay early on Monday morning and Kay was OK with that. The outgoing tide helped us down the Hawkesbury, and we shot past Barrenjoey and out to sea. The nor'-easter was already in, and we made a fast passage back to Sydney Harbour, back to the Balmain Marina, beating the southerly by about an hour. We got 'Dorado' settled into her berth, and I was just taking a moment to relax. We got back just in time,' I said.

'We did well,' Kay answered.

'I've had a marvellous weekend, thank you very much. Can we do it again sometime?'

Sure, I said. We'd organise it.

'Great,' she said. 'I'm going back on board now to tidy up. Give me 20 minutes or so then come across, and we'll have a cup of tea.'

'I'll look forward to it.' A kiss, just a touch on the cheek, and she was gone. So, I got myself tidied up, and cleaned up as far as possible, got some fresh gear on, and went over for my cup of tea. One thing led to another, and I ended up being there for quite

Chapter 3 - Dorado

some time. This is better than going to work, I thought as I blew another shift. It ended when a call came through asking her to go to work urgently. I offered to run her over to Mascot, but she had her own car. She was off to London this time and would be seeing Steve, no doubt. I wished her 'Good Luck' as we parted. I'd better go to work today, I thought, they'll be thinking I've gone missing.

I'd become good friends with Frank McAlister. We'd go to the pub and have long conversations about boats and sailing. Frank was a bit older than me. He'd done 10 years in the Royal Navy before coming to Australia. He worked for the Post Office, as it was then, driving a truck, delivering bulk mail to country areas of NSW. Along with his brother, he'd built his yacht himself. It was of steel construction, a 40ft (12m) ketch and would have been capable of sailing anywhere in the world. Sadly, Frank wasn't confident enough to take it on. But he was interested in coming sailing with me. George Howie's company had moved him to Perth, in Western Australia, so he wasn't going to be around.

I was thinking about a longer trip overseas to a foreign port, maybe Noumea in New Caledonia, about 1,100nm north-east of Sydney. That would be a good test of navigation and seamanship. Eleven days up and eleven days back, with a few days in between for R&R. New Caledonia was a French overseas territory, which would add a bit of spice to the visit. I talked it over with Frank and he was quite keen on the idea. We'd have to do it in May, after the cyclone season, and before winter set in.

That was still a few months away, and there were things to do in the meantime. First of all, I had to get back to work. I'd been slacking off a bit recently, and that couldn't continue. In the longer term, I wanted to circumnavigate the globe with 'Dorado', and that would require lots of money. I did long spells of six days per week. The cab game is like any other business in that you have your good times, and you have your quiet times. So, you have to make the most of good times, and just hang in there when business goes quiet, and it is a bit of a struggle. Some days you can be lucky, and other days the exact opposite. With experience and cunning, you can make a good day out of it, and that is how the best drivers prosper, even on the quiet days.

On the sailing days, I'd been experimenting with 'Dorado's' steering. On long voyages, the single hander must leave the tiller at times for rest, navigation, meals, sail handling etc. I'd read extensively of other people's experiences, and I wanted to see if I could get 'Dorado' to balance and hold a course with nobody on the tiller. In light-moderate conditions, she would hold a course for a while, by trimming the sails just so, and by tying off the tiller with just a touch of weather helm. This worked reasonably well in breezes up to 15 knots as long as the sea conditions remained relatively calm. But when the breeze got up over 15 knots and the seas started to build, control of the steering became much more difficult, and balancing became impossible.

It all came down to the design, which is always a compromise, and very few boats can claim to be anywhere near perfect. Dorado's designer, who was Australian, had borrowed from the well- known Dutch designer, E G van de Stadt, who designed and built plywood yachts of around 23-24ft yachts to race as a class. This design featured a hull with a hard chine. Most hulls are rounded from the deck down to the keel, but chine hulls come straight down to below the waterline, then take an angle of around 135 degrees straight down to the hull. This meant that the underside of the hull was quite flat, and that meant much faster speeds when you were sailing on a reach.

This was true for 'Dorado' and reaching could be quite exciting at times. Going to windward was her weakest point of sailing, and it was hard work getting the best out of her. As the breeze increased going to windward, the yacht would heel over to the point where the chine would start to come out of the water, and that was the point where 'Dorado' got hard to handle. Weather helm increased, and the helmsman needed strong arms to keep her on course. Easing the mainsail helped, but that meant that a hand was required on the mainsheet, and sometimes it was

Chapter 3 - Dorado

better to put the first reef into the main. It was also the case that when we were reaching in a fresh breeze, and the seas were starting to build, 'Dorado' would sometimes broach (shoot across the face of a wave like a surfboard and dive back up into the wind). She'd be just about unmanageable in these situations, and that could be potentially dangerous.

I was looking at these things through the eyes of a single-hander and having a self- steering gear to control the boat at all times. The fact was that I had a lot to learn about sail trimming, and sail management, and when I did eventually get self-steering gear, it must be able to control the boat at all times.

Borowitz, 'Dorado's' designer, had borrowed design ideas from van de Stadt meant for a 22'-23'yacht of plywood construction and applied them to a much bigger (35') yacht of steel construction. He called the design the 'Temptress', and quite a few were built in Australia. John Lake told me that there were quite a few around Hobart, and they raced as a class down there.

Kay had returned from her trip to London, and the big news was that Steve was coming to Sydney in a couple of weeks' time. Kay was going to give up her job as a hostess, and they were going to be married, and settle down. Well, they'd been an item since they were teenagers, and so everybody agreed they should get married; they just took a rather different path, that was all. I met Steve, and he was a very nice guy. He thanked me for looking after Kay, and they did a Sunday sail with me one time. They were moving to WA, where they knew people. They got married over there, and the last I heard they had a couple of kids.

When Kay and Steve left Balmain, they sold the houseboat to Kay's sister, Sue. A lot of people thought Sue and I might make a go of it, but there was no chance of that. We checked each other out and didn't like what we saw. Sue was Kay's younger sister and had been one of twins. Her twin sister had died at birth, and that affected Sue. She worked in the IT industry and socialised at the pub every night with the crowd and got into marijuana big time. She did come sailing with me a couple of times but got violently seasick when I tried to go out of the heads. She would have nothing to do with boats after that, and it was obvious that we had very little in common.

May came around quickly enough, and Frank and I were getting ready for our trip to Noumea. A third crewman was coming with us. Jimmy Peacock, a Liverpool lad, and a friend of Frank from Summer Hill. I'd met Jimmy at the Summer Hill pub, and he seemed a nice enough guy. We were going to a foreign port this time, and that meant we had to have passports, and check out in the same way a ship would.

We completed the formalities and passed through Sydney Heads with a morning westerly breeze to take us on our way. The westerly died out and was replaced by a light nor'easter. At this time of year, the nor'east sea breeze doesn't last, and the breeze soon shifted to nor'-west and freshened up. The barometer was falling, and a change was coming. There had been no bad weather forecast when we checked before setting off, and the breeze was favourable, so we kept going. Our system was two hours on the tiller and 4hrs off. I did the navigation and kept the log up to date. A cup of tea or coffee at every changeover, and food for whoever felt like it. The breeze held at 20-25 knots, and we made great progress.

By late on the second day, we were more than halfway to Lord Howe Island. Our course to Noumea passed fairly close to Lord Howe, and it would be good if we could sight it as a navigation check. About 90nm north of Lord Howe lies Middleton Reef, which is just a large coral reef, and a potential danger for the unwary sailor. I wanted to be sure that we kept well away from that danger.

We were now getting forecasts of a change in the weather; the wind was going to back from west to sou'-west to south to sou'-east. A low was to form in the south Tasman Sea, and a preliminary gale warning was issued. The wind with us was sou'-west and freshening, and a few hours later it backed to the south, and cloud started to form. Later, it started to rain. I'd gotten some morning sights before the cloud closed in, and I was reasonably confident of our position.

Chapter 3 - Dorado

There was some talk that maybe we could stop at Lord Howe and wait out the gale. Too dangerous to try and approach in this weather, I thought; much safer to keep going. Late in the afternoon the clouds partially cleared, and we saw the misty outline of the northern coast of Lord Howe, and that gave me the navigation check that I wanted. The wind was now sou'-east, and gusting up to 40 knots, and the seas were building up. We were a long way north of the low, and so we could expect that conditions might slowly improve from here.

During the night, there was a bit of drama. I was resting on my bunk when suddenly I got thrown out on the floor, and gallons of water came rushing over me. I rushed up on deck to find Jimmy floundering around up to his waist in water. I think what had happened was that 'Dorado' had gotten away from Jimmy and broached in front of a breaking wave. He shouldn't have been up there on his own in these conditions; there should have been two hands, at least. I quickly got 'Dorado' sorted out and back on course, there didn't seem to be any damage that I could see, anyway.

The gale force conditions eased slowly over the next four days, but the skies remained cloudy, and I got no sights, so we were relying on our dead reckoning as we approached New Caledonia. Our course from Lord Howe was around nor'-nor'-east and the wind was from the sou'east, and that meant we were on a beam reach all the way. There would have been a bit of leeway; hard to judge how much, but we were starting to approach the coast, that was for sure.

It was a bit of a nervous night. The reef in this part of New Caledonia stands well out from the coast, and we didn't want to make any mistakes. The breeze had died away to about five knots, and the sea was almost calm. Next morning, still nothing to see, still overcast. We keep going slowly, slowly, then in the early afternoon we see land, then we see the reef, then we see a passage through the reef, and land behind.

I knew straight away that it was not the entrance to the main channel up to Noumea. The Amédée lighthouse dominates that entrance, but there is another passage through the reef, and that also gives access to Noumea. Maybe that's where we were. We went through the pass to explore but we didn't go very far. There were coral heads everywhere. We were obviously in the wrong place. The coral heads were crowding in on us, we had to get out of there.

Then a stroke of luck; after five days of cloudy skies, the sun came out. I grabbed the sextant and got busy. It wasn't long before I had us pinpointed. We were about 40nm up the coast from the Amédée light. OK, the weather is calm, the sea is calm, we're motoring along at about four knots. We'll follow the reef down and be at Amédée about dawn.

It was late afternoon as we motored along not far from the reef. We'll have to go out to sea a bit overnight, I thought; it's too dangerous being this close to the reef. Then we saw a ship, a bit farther out than us, and it didn't seem to be travelling very fast. We'd better let that go past before we head out. I kept watching it. 'It doesn't seem to be moving,' I said to Frank.

'No, it doesn't have any lights on, either' he replied.

Something was wrong; alarm bells were ringing. What are we going to do? Crunch. It was what I had feared. The reef had bulged out and, in fact, the ship we had been looking at was stranded on the reef. Where we hit, the reef was actually three or four feet (1-1.25m) below sea level. The calm conditions, and the fading light, and there'd been no breaking waves to give it away. We were just starting to realise that something was wrong when…

Too late!! We struggled to get off, but a rising tide running over the reef pushed us farther on. There was a lot of scraping and grinding coming from the bottom of the keel, until eventually we were right over the reef, and into deep water. It was fully dark now, and there was nothing we could do right now. I went below looking to see whether we were taking any water. Nothing. All that crunching and grinding, but no leaks. The bottom of the keel is a large flat plate 3" thick, part of the ballast, and that had saved us.

Chapter 3 - Dorado

We put the anchor out to try and hold our position; there was no wind, but there might be a little bit of current. We took turns to be on watch, but nothing happened. Everything remained calm. We were very lucky with the weather, any kind of north or west wind, and we would have seas breaking over the reef. It just doesn't bear thinking about.

In the morning, the tide had gone out. There wasn't enough water to float 'Dorado' and she was heeled over at a steep angle. I launched the dinghy and went to have a look around. We were in a pool surrounded by coral. There was one place where we might get out at high tide, and pass through to another pool, maybe at high tide, but it was very narrow.

Frank was calling out to me. He could see people on the ship; somebody was getting into an inflatable dinghy and coming our way. Bonjour! We soon worked out it would be better if we spoke English. The guy was the boss of a team of guys who were cutting up the ship for scrap. We saw you there, he said, and you look like you're in some kind of trouble, so I've come over to see if I can help. We discussed it for a while and then he said, 'Look, I'm going ashore for a couple of days, and if you want to come with me, I can introduce you to someone who has tugs that might be able to tow you off.

'You can come with me now,' he said.

Spur of the moment decisions. What do you think the weather is going to do, I asked him?

'The weather will be fine for at least a few days,' he said.

I looked at Frank and Jimmy. What do you reckon, guys? Do you mind if I go and see what can be done?

OK, they said, but I could see misgivings in their eyes.

'I am coming back,' I said. 'There is no way I'm going to abandon the boat here.' 'OK, OK!'

We were a fair distance offshore, and it took some time to get there. It would be a very long way indeed if you were trying to row a dinghy. He had a vehicle parked there, and it was about an hour's drive into Noumea. He dropped me at a hotel. They're expecting you, he said. Go and check in and freshen up, and I'll come back in an hour. I walked in the door. There were several people there, and they were reporters, and they wanted me. I was taken aback. What could I tell them, anyway? They already knew what had happened.

The penny was starting to drop. I was being set up. I bailed out of there as soon as I could and got to my room and had a shower and freshened up. When I went back to the foyer, my friend was there waiting for me. We got in his vehicle and drove down to the harbour to the office of the tug company.

We went to the tug master's office and sat down and had a discussion.

Yes, he had a small tug available. It would take three days to do the job. One day up there. Set up to tow us off at high tide at the place we came over. Hopefully no damage to the boat, and one day back to Noumea. Somewhere in the conversation the word salvage was mentioned. That rang a bell, and I remembered a conversation with Alf about salvage rules. Not Good. They were talking about money now, and it sounded like telephone numbers.

'In $A?' I asked. It still sounded like telephone numbers. Then the question was asked, did I want to go ahead with it?

'No, not right now. I need more time to think about it.

'Thanks very much for your help... I appreciate what you say... It's a difficult time for me.'

He dropped me off at the hotel. I'll come back and pick you up at 7am he said. Don't have any breakfast; we'll have breakfast on the way. And we did. It was a restaurant serving breakfast. The tables were all occupied, but places had been reserved for us. He really was trying to impress me. Lashings of bacon and eggs, and beautiful French bread, and wine, too... Well, they would, wouldn't they!

After a couple of glasses of wine, he popped the question. Would I be interested in selling the boat. No that'd be the last resort, I told him. I wanted to go back there and then, and reconnoitre, to see if there was any way we could get out of there; maybe even try to get back over the reef the way we came in, at

Chapter 3 - Dorado

high tide. But if the weather turned nasty, with breaking waves, etc, and there was a threat to life and limb, and we had to abandon the boat well…

'There's a high tide tomorrow morning,' he said, 'higher than the one that put you in there. My work boat has a strong engine. Maybe I can tow you with the work boat and get you out of there.'

Well, that's an idea, I thought, but in the back of my mind a little voice was saying, 'A tow from his boat is an act of salvage'.

We got back out to the reef. It was low tide and 'Dorado' looked pathetic, heeled over on her side. The guys were sitting on the high side of the deck. Frank yelled out as we approached.

'We've found a way out! It's easy!'

'So, do you want me to come over tomorrow morning,' the salvage boss said.

'Look,' I said, 'you've been very good to me. I appreciate what you've done, and I thank you for what you've done. I think we'll be ok here now. I hope so anyway.

'Bonne chance, he yelled as he revved his outboard and took off.

You should've stayed here, said Frank, we could have been outta here by now.

That's easy to say now, I said, but no harm has come from my going yesterday.

'They came over here in a plane yesterday,' says Jimmy. 'Probably taking photos.'

'That's right,' I said. 'Look…' and I showed them a copy of the local newspaper, which I'd picked up at the hotel. There we were all over the front page. 'C'est le Dorado sur la Récif,' screamed the headline, and a photo to go.

After a cup of tea and some French bread that I'd brought for them, Frank and I got in the dinghy and went to check out our exit route. The first gap was the trickiest. I'd seen it yesterday, and it looked narrow with some sharp edges. While I was away Frank had gotten out with the big hammer and had opened it out quite a bit. It now looked quite passable at high tide. That led us into a second pool, which also had a gap, that led us into a third pool which had a wide gap which took us into a channel and back out to sea. Ye Gods! How amazing is that! When's the next high tide? The biggest tide is tomorrow morning at about 7am. It is in daylight hours, and we need that to see where we are going. That's it, then.

When the time came, we sent Frank out in the dinghy to lead the way. The weather remained fine throughout, and we had to use the auxiliary motor to get us all the way to Noumea. We went straight to the yacht club, which was recommended at that time. They arranged a berth for us, and called the customs, who came out and completed formalities.

We were waiting for the officials to come, and we saw a vehicle arrive at the club. Out stepped a man wearing what looked like a gaberdine raincoat, with a slouch hat, and a pipe in his mouth. 'Who does that remind you of?' I said to Frank.

'Maigret,' he said, instantly.

'Maigret' was a very decent guy, and the formalities were soon completed, and we could go into town to have a look around. I was to meet 'Maigret' a number of times in my sailing years, and he always looked after me.

A strong burst of trade winds came through while we were in Noumea, reminding us that it was time to get going. Frank and Jimmy had taken a month off work, and we had a trip ahead of us that would take at least 10 days. The trades would give us a start, then we would almost certainly encounter westerlies. From Noumea to the Amédée is straight south for 20nm, so we were close hauled for the start of the voyage. From Amédée, the course was west. We chose to go west because we'd get the advantage of an ocean current all the way down to the coast where the current splits. The main part of it turns sou'-west and becomes the East Australian Current, and again that is an advantage to us.

And that's how it worked out. After the first day, the trade wind eased, and we were able to hoist the spinnaker. In fact, the spinnaker stayed up for seven days, by which time we'd crossed the Coral Sea, picked up the coast, and held the breeze (now a

Chapter 3 - Dorado

nor'-easter) all the way down the coast until we'd past Crowdy Head and were approaching Port Stephens.

They'd been forecasting a change for a couple of days, and we were kind of hoping it might hold off for another day. Not to be. The sky was building up, and we weren't surprised when they forecast a thunderstorm. We dropped all sail while the storm passed through, which took about an hour. The breeze had backed around to the west and was only moderate to start with but increased rapidly and backed further to the sou'-west at 25 knots.

We were making good progress with a reef in the main and #3 jib when the wind backed again to south 30 knots. The seas were starting to build up now and we had to take another reef in the main. Now we were getting forced down on the coast, but there was still plenty of clearance, and we held on hoping for another wind change. This came when it backed to sou'-east 25 knots. We could lay Sydney Heads now, and as the dawn came up, we were passing Barrenjoey. The wind eased more and went east 15 knots, and we passed through the heads in the afternoon, called in at the Customs at Neutral Bay, completed formalities, and got back to Balmain before dark.

The main lesson learned on this trip was that dead reckoning is important. The conditions were rough, and I should have made more allowance for leeway. Also, I learnt about the current that runs to the west caused by the trade winds. I learnt that I must record progress every hour, at least: course and distance travelled, wind direction and strength, sea conditions, barometer, sightings made of land or other vessels, and anything else that might be relevant to navigation. People say that I should have had an RDF (radio direction finder), but I'd heard mixed reports about the hand-held versions that people have on yachts. Perhaps one might have helped us get to Noumea without running aground. I was not a believer in them, and I never had one.

I was still hoping that I might find a locally made self-steering gear. The time was coming when I was going to have to make a decision to get one of the top-rated British versions (Hasler or Aries) or find a locally designed one that could do the job. Anyway, it was back to work for me for the time being.

My first shift back, and I get over to the 'Golden Fleece' at Drummoyne for my 3pm start. I've been away for a month, and I don't even know what car I'm driving. I went to see the manager, Bob Burgess, and he told me the car to drive. 'It'll be in shortly,' he said. 'You'll get a surprise when it comes in.'

Oh really! So, I get out on the driveway, and have a chat with the other guys about to start work and, eventually, I see the car coming up the driveway. Oh good, I think, here it comes now. Who's that driving? The car pulls up at the petrol pump, the door opens, and the driver jumps out looking me right in the eye. 'Hi,' she says. 'You must be George.'

'Bob told me I'd get a surprise,' I said. 'I must say, I wasn't expecting such a pleasant surprise.'

Her name was Jenny, and we became friends. She'd pick me up on the way to work; we went out to dinner; I stayed at her place, and she came sailing with me, and we went to Refuge Bay and enjoyed that. Jenny was a country girl who had come to the city to become a model. She joined one of the top troupes, and had a successful career living the high life, as they do. Then she had the misfortune to be involved in a traffic accident and had a minor injury to her face. It was minor, but it was enough to affect her career, and it wasn't long before she was out of it altogether.

Jenny had then been faced with keeping body and soul together, so she decided to become a cabdriver. Being a cabdriver is a tough life for the guys, and very few last more than a few shifts. Not many women even think of trying it and, of those who do, none last very long. Jenny was an attractive person with a nice personality, and I'm sure she would do well in a public relations situation. We did one last trip, up to Port Stephens this time, which was nice, but we did get some nasty weather on the way back, which she didn't enjoy too much. I was fairly certain that

Chapter 3 - Dorado

there was at least one other guy around at that time, and it was no surprise to me when she called it quits.

George Howie was back in Sydney, and we had a few get togethers. He knew that I wanted to circumnavigate, but I felt that I wanted one long trip single-handed under my belt before I set off on that one, and the self-steering gear situation hadn't been resolved yet. It was George who made the suggestion. 'Look,' he said, 'I'm due for some holidays next April/May. Why don't we plan a trip to, say, Fiji. I'll sail up with you, and you can single-hand it back.' George said he'd get a workmate, Dave Mitchell, to come with us. It would be a great experience for him, too, he reckoned.

Well, that was next year. What I had to do first was to keep working to earn the $$$s that I was going to need for the circumnavigation, and also continue to experiment with sail management. I knew that I tended to haul the mainsail too hard when going to windward, and if I eased it a little the boat would perform better. No wind is continuously stable in force or direction. It will gust in from one direction, and then slowly ease and veer in direction over a period of time. I also found that it was much better to put the first reef in the mainsail just a little early in an increasing breeze. It made keeping control a lot easier. On a racing boat there will be a crew member who is continuously in charge of the mainsail, and he/she would adjust the sheet as required, and the same for the genoa. A single hander does not have the luxury of a crew but must still be aware that things need adjustment to get the best out of the boat. I discovered after I got the Aries (self-steering gear) that it needed regular attention, also. It is not a set and forget piece of equipment and needed regular attention and, maybe, adjustment to get the best out of it. Cruising in a yacht may not be a race, but the sailor is duty bound to get the best out of his/her boat and arrive at the destination as soon as possible.

It was around this time that I met Marise. I was working a Monday night. It had been very quiet, even for a Monday, but it was coming to that time of night; the pubs were coming out, the movies were coming out, there'd be a few fares around in the city. I was coming down Parramatta Road, just where it widens out and becomes Broadway.

There are a few shops and businesses there, and one of them is 'The Kings Court', which is a massage parlour. It's a bit more than just a massage parlour, according to some of my clients, who tell me that if you're nice to them, and offer a few extra $$$$s, you can get a 'full service'. I'm just approaching the Broadway when I see someone outside 'The Kings Court' and she's flagging me down. An attractive looking gal'; she opens the door and gets in the front seat. That's a good sign. She might be on for a chat anyway.

'Hello! Where would you like to go to tonight.'

'Home!' she said.

'What? Your place or mine?'

Oh well, it's good for a laugh, isn't it. Lidcombe, she said. OK,

Lidcombe, out in the western suburbs. That's a good fare on a quiet Monday night. We chatted away as we travelled. Business is quiet; same problem for both of us. As we approached Lidcombe, she said to me. 'George, this has been a very quiet night for me, and I haven't got the money to pay you, but if you come in with me, we could have a cup of tea.'

'So, what are you saying exactly? Have you got money at your place?' I said.

'Better than that,' she said, a small smile playing around her lips; better than that… So, I took the chance and went in, and it worked out well. I picked her up the next week and the week after that. We had an evening out one night, and one day she came down to the boat. She wasn't interested in boats, and definitely didn't want to come for a sail. The next Monday I went to pick her up she wasn't there. I went round the back entrance in the laneway and saw the bouncer. Is your name George, he says? Yes. She's busy with a client, he says; don't wait for her.

'By the way,' he says, 'I've got these people here, and they're just about to ring for a cab… D'you wanna take them?'

Chapter 3 - Dorado

'Yeah, sure mate; no problem.'

Frank came over to see me one day. It was unusual for him to come over to our side, so it wasn't just a casual visit.

'Do you remember John Vago?' he said.

'I never met him,' I replied, 'but I've heard of him'.

He came down to see me the other day, says Frank, and we had a bit of a chat. 'He's been approached by some rich Americans who want to go sailing. He's got a boat, but he needs an experienced crew.

'It's easy mooney man, easy mooney.'

Frank was a Liverpool lad, of course, and his accent made money into mooney. I'm going down to see him next week, says Frank. D'you wanna come with me? OK, I said.

I was already sceptical about the rich Americans, but Frank was a friend, and I thought, well, there's no harm to it. So next week we take a trip down to Cronulla, in the Port Hacking area of southern Sydney, and we met John Vago, with his wife and kids. Vago was a Hungarian ne'er-dowell who lived off government benefits in a Housing Commission flat. He was always full of big talk, and plans, and ideas. He must have had money at one time, because he had a yacht in a marina berth at Cronulla near the railway station. We went down to see the yacht. It was a 50ft ferro-cement hull, with mast and rigging. Below, there was a very basic fit out with a big auxiliary in position and presumably working.

'Where's all the sails? I asked Vago.

'Oh! We've just got a mainsail at the moment,' he said. 'We're getting some more sails soon.'

'So where are the rich Americans?' I asked.

'They're not going to be sailing with us,' he said. 'We're delivering some goods to them.'

Smuggling, I thought; say no more for now.

The whole scenario unfolded slowly. This was 1974, and drug smuggling hadn't really got going yet; but there were other things: rare birds, for instance. Rare birds were protected by federal law, but the government had seen fit to licence certain parties, which allowed them to sell birds overseas. Bird specialists who had missed out on a licence were upset by this and set up their own (illegal) trade. A matched pair of the right species of bird could fetch tens of thousands of $$$s landed on the west coast of the USA. So, there was plenty of incentive. The hard part was getting the birds and getting them out of Australia. To this end they had a guy — call him an 'enabler' — in Sydney who knew the local scene, and who to see etc. He'd found this guy, an ex-policeman, who was a bird fancier, and who knew how to get the required birds. I'm not sure how they got in touch with John Vago, but his boat was just about perfect for their purpose.

We came down a second time and met the ex-cop. He explained to us that he was licensed to carry a gun, and he showed it to us. He introduced us to a friend of his who would be coming with us, he said. He was a big guy, a tough guy. He'd worked on the fishing boats and knew all about diesel engines. Vago told us later that he was a standover man and had done time for violent assault.

Another time, we went down, and Frank took a couple of jibs off his own boat, and also the compass. 'Frank, you're doing the wrong thing,' I told him. But Frank wanted the mooney he thought he was going to get and wasn't listening. We set the compass up on Vago's boat, but it was no good. It needed to be set up professionally and adjusted. Nobody knew what to do, so I foolishly gave them the number of the guy who set up the the compass on 'Dorado', Captain Bozier. Because my name was mentioned he was good enough to go all the way down to Cronulla and set their compass up, and then they didn't pay him. I found this out from Frank in the next week and went over myself to his office at Crow's Nest and paid the money. It was a bad look, though, and the relationship was never the same thereafter. I was disappointed in Frank; he wanted to do this trip and get the mooney. I wasn't happy at all. I could see a disaster coming, and I was looking for a way out.

Chapter 3 - Dorado

We had one last trip down to Cronulla, and the ex-cop explained what was going to happen. The birds would be ready in two weeks' time, and they'd be brought down at night and loaded on to the yacht. We were to leave immediately, and head out to sea and be well clear of the coast by dawn. We would be heading into to New Zealand waters, and rendezvous with a Kiwi trawler about 50nm off New Plymouth, on the west coast of NZ's North Island. We would transfer the birds, then head back to Australia.

That sounded nice and simple, but I knew there were all sorts of potential problems with it. I said nothing, but I now knew that I wasn't going to go. He was still talking. I want you to bring your passports down and give them to John next week. 'Passports? What do we need passports for?' I said. Silence.

'We're not entering NZ, are we? Just dropping stuff off out to sea, and then coming back.'

'Nobody needs to know where we've been. Just bring your passport, OK!'

There was no mistaking the menace in his voice.

'Yeah, right oh, no problem. I hear what you say.' 'No passport and you won't be going, d'you understand?'

Sure.

I got stuck into Frank on the way back to Balmain, trying to dissuade him, but he wasn't having any of it. He was going, he was gonna get his mooney and that was it. I'd been mates with Frank for the last three to four years and I thought I knew him well. Not as well as I thought, obviously. The trip was set up to fail; the yacht was poorly equipped for an ocean voyage and didn't even have a life raft. Vago was a lightweight with very little or no offshore experience. The ex-cop and his offsider were hard cases involved in a criminal enterprise and would stop at nothing to get it done. The idea of going away on an ocean voyage with these two did not appeal at all. It was downright dangerous. Frank was the only one with any real knowledge of sailing and navigating, and he wouldn't get much help from any of them.

I told Frank that I wasn't going to go. I wasn't going to take time off. I was going to keep working. I had other things to do. Just let them know the next time that you see them, I said.

As it happened, they were held up by to the unavailability of some of the birds and didn't get going for another month. Frank just disappeared from Balmain one day and I knew they must have gone. I saw or heard nothing more for about six weeks when Frank reappeared on the Marina. I wanted to hear the story, of course, but I had to wait until one night when we went up the pub for a few beers. I'd estimated that if the trip had been a success, they would have been back in not much more than a month, but six weeks was too long, and the sheepish look on his face told me that there had been drama. The truck delivering the birds had arrived on time in the middle of the night, and were promptly loaded on board, he told me. They left the marina immediately and set off.

They hadn't got very far before they'd run aground on the bar in Port

Hacking. It was low tide; surely someone should have known that. Port Hacking bar can be very dangerous in strong sou'-east weather and gets closed altogether at times. The weather that morning was calm, but you should always allow yourself a bit of tide on that bar. Was that a bad omen?

They got over the bar, and the voyage proceeded uneventfully until they began to approach New Plymouth when the westerly wind started to freshen up, and a gale warning was issued. Their means of contact was by CB radio, and after a few attempts they made contact with their friend in New Plymouth. Can't come out in this weather, was the message, heave to until conditions ease. They had to move away from the coast and heave to. Twenty-four hours later, conditions had eased a bit and they made contact again. This time the message was different. 'The authorities know there is a yacht out there, and they want to know who it is and what they're doing. Recommend that you abort your mission.' You could imagine the consternation aboard the yacht… What were they going to do?

Soon after that, an aircraft appeared and flew around them.

Chapter 3 - Dorado

They put their CB onto the emergency frequency, and sure enough there was a message for them. 'This is the NZ Customs Service. You must proceed directly to New Plymouth harbour. We are waiting for you there.'

What were the options? No options really; they were going to have to go to New Plymouth. But first they were going to have to get rid of the birds. After the plane had gone, they brought all the cages up on deck, and let the birds go, then dumped the cages over the side. Then there was a big clean up below decks, so every trace of the birds was gone.

They were arrested as soon as they entered New Plymouth harbour and held for several days. In the end, the Kiwis imposed only token penalties on them, but the NZ Customs had been in touch with the Australian Customs, and the yacht was ordered to go straight back to Sydney harbour and surrender. On arrival in Sydney, the crew were re-arrested, and faced court proceedings on criminal charges, and the yacht was confiscated. All the crew received heavy fines.

Frank was lucky not to lose his job, but he did lose his compass and sails, and other gear he had taken on board. He was quite bitter about it all, and we were never as close again as we had been. The yacht was eventually auctioned by Customs, and Vago went and put in a bid for it. Unsuccessfully!

It was a dark time for me. I didn't go on the trip, but somehow or other Customs got hold of my name, and I became a person of interest. I'd lost the way for a while, sailing too close to the wind.

I was reading a local sailing magazine one day, and I saw an advert for self-steering gear. Local manufacture, strong control system, it said. I'd been thinking of getting an Aries, from the UK, but the cost was putting me off. Was this local offering any good? Would it do the job? I contacted the people and went and bought one. This was the last chance; if this couldn't do the job, it would have to be the Aries. I can't remember what it cost, but it was only a fraction of the Aries cost. The main feature of the gear was a vane made of lightweight fibreglass. It was about two feet long and shaped as an aerofoil. There were two lead counterweights to keep the vane upright, and a system of lines leading to the tiller, which steered the boat.

I soon had it mounted on the stern of 'Dorado' and got out on the water to give it a try. It was no good trying it in the harbour with all the other traffic around. We had to be offshore where we had a steady sea breeze of 15 knots, and a slight swell from the sou'-east. It worked reasonably well to windward with lots of adjustments to sails and gear, but then that's what sailing is all about isn't it? On a reach, not quite so good… but this is just the first time. Lots more testing needed. On a run, 'Dorado' yaws about all over the place, and there is danger of an accidental jibe. That could be dangerous.

There are still a few weeks till the Fiji trip, so I'll get in a few more trial sails before then. And I did, and I discovered that this gear was not the complete answer. I could leave it for a few minutes to go and take a navigational sight, or go below to make a cup of coffee, but I couldn't go below and lay down and rest; the boat would wander off course or worse.

Anyway, I hadn't tried it in strong weather yet. The Fiji trip was coming soon, and there were going to be three of us, and we'd steer by hand doing shifts as we'd done before on the way up there. But I was going to do the return trip single handed, and that would be the time for finding out if this self-steering gear was up to a circumnavigation.

It was clear that someone had given Customs my name when I went to do the clearance for the Fiji trip. I was questioned about my role in the NZ escapade. I just said that I was a friend of Frank, and I had tried to talk him out of being involved in it. I sensed that it was going to be a disaster, and I tried to tell him to have nothing to do with it, but he wouldn't listen. They went on asking me questions, but there was nothing I could tell them that they didn't know already. They gave me clearance for the boat and the crew but told me to stop by (Customs in) Neutral Bay

Chapter 3 - Dorado

tomorrow and pick up the paperwork. We did that; they had the paperwork ready. They also came on board and made an inspection, which they are entitled to do. Dave Mitchell took exception to that and got a ticking off for his trouble. I was studiously looking away. Then it was down the harbour, and out to sea, followed by a Customs launch. They were seeing us off the premises, I told Dave when he commented on it.

We soon settled into the routine. Fiji is about 2,200 nautical miles from Sydney, and we covered it in 21 days. We had a variety of weather, no gales, no days without sun sights. My navigation was much improved. At dawn on the 21st day, the island of Kadavu, to the south of Fiji's main island, Viti Levu, with the nation's capital, Suva, was in sight, and we shot through the pass into Suva harbour a few hours later.

The Customs were nice to us, and formalities were soon complete. The Customs guys told us that the best place to anchor was nearby at a place they referred to as 'The Bay of Islands', where there was a tourist hotel. Looking at Google Earth now, I can identify the place quite clearly. It's called the Novotel, although that was not the name back then. Ah well, I suppose it was nearly 50 years ago. The hotel made us welcome and allowed us to use their facilities.

After 21 days at sea, the guys were ready to play up, and the 'Whisky au Go Go' got big licks. Suva was a duty-free port then, and Alf had given me instructions on who to go and see to get the best prices. I wanted to get a good quality camera, and a new watch. To that end, I bought a Minolta camera and a Citizen watch. The camera was great and stayed with me all the way to Cape Horn. It rests today at the bottom of the cliffs together with thousands of photos. The Citizen watch was to stay secure in its box. It was going to be my chronometer, and only to be used when taking sights. I had a Seiko watch on my wrist, which had served me well for a number of years. The second hand was a bit hard to read at times, but it would still be a good back up.

The guys had their flight booked for three days after we arrived in Suva, and all of a sudden it was time to go. I was going to stay in Fiji for a while. Alf had given me some contacts in Lautoka, and I was going to cruise around there, and then perhaps up to the Yasawa Islands, to the nor'-west of Viti Levu.

I took it easy, just day sailing, and anchoring overnight. My first stop was at Beqa, the island of the firewalkers. I'd seen them perform their act whilst we were in Suva. My next move was to Sanasana, which is a resort set on a small bay which provides a safe anchorage. Nowadays, it's better known as Natadola. It's a beautiful beach with a number of resorts of varying standards clustered around it. I'm not sure whether I was welcome, but nobody said anything when I walked up to the bar and had a beer. Next day, I intended to get to Lautoka, but it was a bit farther than I thought, and I stopped at Port Devatran which is also a resort with a nice safe anchorage. Again, I took a walk ashore and found the bar, where I met a guy with his family who invited me to dinner. That was a bit of a challenge for me, and I had to go back on board and smarten myself up before meeting these generous people and enjoying a lovely meal.

Next day, I completed the trip to Lautoka and tied up at the wharf around lunchtime. There were no other yachts there, but I could see some 'Blue Lagoon' cruise vessels there. I remembered what Alf had told me and decided to go and see them the next day. In the meantime, I took a walk into town to have a look around. Lautoka is the second largest town in Fiji and dominates the western side of Viti Levu. The main business in Fiji is tourism, of course, but they also produce quite a lot of sugar, and coconuts (copra).

The sugar industry had been started by the British in colonial times, and straight away they had a problem: the native Fijians flat out refused to work for the British. That didn't faze the British at all; they brought in Indian labour, as they had done in South Africa, and established a viable industry. The Indian population grew, and at the time that I was there they were about equal with the native Fijians. The two races didn't like or trust each other, but there was very little of the violence that you see in some parts of the world. The Fijian army, which consisted entirely

Chapter 3 - Dorado

of native Fijians, ran the country, and are well known for their work as peacekeepers, especially in the Pacific. In later years there was a democratic election which was narrowly won by the Indian candidates who appointed an Indian prime minister. The Army wasn't having that though, and took over the parliament, and dismissed all the politicians. Fiji is our country they said, and we run it! And it is still that way today.

There is another outlet for the big, brawny, aggressive Fijian guys in the game of rugby, and they are very good at it, and have dominated the sevens game for some years now.

It didn't take long to look around Lautoka; the only thing of interest was a pub, and I went in to have a beer. I wasn't very impressed. There were a lot of dodgy-looking characters eyeing me off. I wouldn't wanna be here later on in the evening, I thought, as I finished off my beer and headed out the door. The wharf was easy to find due to a big threestorey building nearby. The ground floor was a fruit and vegetable market; on the second floor was a Chinese restaurant; and on the top floor was 'Wilsons Nite Club'. Alf had warned me about Wilson's before I left Sydney. 'Don't go there,' he'd told me. 'It's dangerous. You go there, and you might not come out alive.' Well, it doesn't look like much, I thought as I walked by.

The next morning, I walked over to the Blue Lagoon wharf, and introduced myself. I met the skipper of one of the cruise boats and we had a chat. He knew Alfie Taylor well, he said. We always have a good time when Alf's around. So, we chatted on for a bit, and I asked him: I was thinking of taking a sail over to the Yasawas, but I don't have the chart of the area. Could I possibly buy a chart from you? Aha, he said, there was no chart of the Yasawa group and visitors were not allowed to go there. The only way you can go there is if you have a local on board who can show you the way. Then he said, it just so happened that a member of his crew was just taking holidays, and his home was on one of the Yasawa islands. He offered to introduce me to him.

And so, I got to meet Setariki Sue Sue. We had a bit of a chat, and decided we could leave that day, but first, we had to go to the markets. The markets were run by Indians, and it was a cut-throat affair. They see a white face coming, and they are all over you like a rash. But not today. Setariki knew where he was going, and he strode through the middle of them right to the guy who sells kava. Kava is a plant that grows around the western Pacific islands. Its dried roots can be pounded into powder and made into a social drink. It's very popular in Fiji and very much part of the culture. Kava is non-alcoholic—it's a mild narcotic—and induces feelings of peace and serenity. We were going visiting in the Yasawas, and it was important to present the village chief with kava to show respect. It was expensive to buy the real stuff—not the packets— but Setariki wasn't paying for it, so only the best would do.

We got going about lunchtime, and headed west, and it was obvious from the start that I needed someone with local knowledge to sail these waters. The whole area was riddled with reefs and coral heads. It was late in the afternoon before we rounded a headland and came to anchor in a pleasant bay in front of a village, which was Setariki's home village. Setariki's girlfriend had come with him, and they went ashore to meet his family. Nothing would happen tonight, Setariki told me. I'll come and get you in the morning. I did go ashore for a bit of a walk for 30 minutes, but then got back on board for the night. In the morning, Setariki came for me. We went to meet Setariki's family, then we went to meet the chief. That's how it's done in Fiji. A stranger should not just walk into a village; they must seek out the chief and go through a formal process known as sevusevu, to ask permission to be in the village and to pay respect and tribute to the chief. The tribute is a packet of kava.

Setariki had broken the kava into packages and the chief got the biggest one. Then I was introduced to the chief and thanked him for allowing me to come to his village. We were invited to come back later to take some kava with the chief and the elders, then we left. They're catching some fish today Setariki said; let's go and watch them.

We walked to a beach on the other side of the Island and

Chapter 3 - Dorado

found a crowd of people. It was high tide, and they were ready to start. About 30-40 of them got into the water and formed a large semi-circle, maybe 200m across. They then started beating the water and edging slowly toward the beach. After a while, it became obvious what was going on: the fish were trapped, and you could see them racing up and down looking for a way out. It went on, probably for an hour or more by which time the fish were trapped in an ever-decreasing pool. That was the cue for the young guys, who'd been standing watching till now. Suddenly, they rushed into the water and started grabbing the fish by hand and throwing them ashore. In no time at all, there was a pile of fish on the beach, easily over a hundred, and some of them quite large.

'Feast tomorrow,' said Setariki, digging me in the ribs with his elbow.

Later in the afternoon, Setariki took me over to a large circular hut where the kava ceremony was about to take place. We all sat down on the ground, cross-legged, and the chief led the proceedings. Two young ladies at the back of the hut had prepared the kava, and there was much talk in Fijian, and ceremonial clapping, which I joined in with as it was passed around. I was given a small cup, filled by the young lady, and I drank it down as I'd seen the others do. It had a very earthy taste, not unpleasant, and almost straight away I felt myself relax, and a feeling of being at peace with the world came over me. The bowl came around again, and I had some more.

I drifted off into a reverie, and quite some time had gone by, when Setariki shook my arm, and got me up. It had been an experience. The effect of the kava lasted for a while, and we didn't do much for the rest of the day. The next morning everybody was buzzing around getting ready for the feast. There was a big open space in the centre of the village and that was where it was going to be. It started in the middle of the day and went on all afternoon. First, we had some fish soup, which was delicious, then came all the Islander veggies, Taro, Dallo, Cassava, Yams, and others I couldn't remember the names of. They are all protein rich and lie heavily on the stomach. I tried to limit them, but Setariki was egging me on to have more. I was brought up on a Scottish diet of porridge, bread, potatoes, turnip, pumpkin etc, but these Islander veggies were too much for me. Then came the fish, most of it barbecued, some apparently eaten raw. I watched how Setariki managed it, and I copied him. It was delicious, spiced with Islander herbs. I'm not a fish eater as a rule, but this was special, and I enjoyed it. During the course of the afternoon someone came along and placed a large dish a couple of feet in front of me. On the dish was a great big fish head. The eye seemed to be watching me. As the afternoon wore on, and things were slowly winding down Setariki grabbed my arm and said, George. You're the guest of honour here, the fish head is for you. I was aghast, already choc a block full, I couldn't face more food of any kind let alone a fish head. Everybody in the circle was watching. I glanced at Setariki, was he joking? He gestured back to me. Just pick it up and put it to your mouth.

I felt I had to do something; it was expected. I reached out and picked up the fish head and brought it up to my mouth, a small piece of flesh was hanging off one corner. My teeth closed on the flesh, and I chewed on it. A great shout went up, and clapping, and laughing. That's fine, says Setariki, you can put it down now, everybody's happy. I put it back on the plate and Setariki grabbed it, and put to his mouth, and crunch, crunch, crunch, the whole thing went down, including the eyes. What a feat, I clapped and cheered along with the rest. It was already late in the afternoon, and people were starting to disperse.

It had been a big day, I needed to lie down and rest. I'd had nothing to drink all day except some flavoured water. There was no alcohol of any kind allowed on the tribal homelands—it's a strictly enforced law. But I had a bottle of rum hidden away on board, and a few cans of coca cola. A couple of good shots of rum and coke were just what I needed to relax and fall asleep.

The next three days were spent going round local islands and villages, where Setariki had friends or acquaintances. Every place we went, we presented the chief with kava, a sign of respect

Chapter 3 - Dorado

that gave us a free run of the village to meet and greet whoever and whatever.

On the fourth day, we visited a place that is a highlight of the Blue

Lagoon cruise. At the sou'-east end of Yasawa Island, in the centre of the Yasawa chain, there is a small bay. Sheer cliffs rise straight out of the water, and there is no landing. But if you dive down into the water about 20ft there is an opening into a cave, and if you follow this cave for about 50ft, and then come to the surface you will find yourself in a deep pool of water. It is partially open at the top, and the sunlight comes in and lights up stalactites that have grown over the centuries, and it all makes for spectacular sightseeing. It was a bit of a challenge for me: could I hold my breath that long? Yes, as it happened, and I enjoyed it very much. Shame my camera is not waterproof.

From Yasawa Island, we headed back to Setariki's home island. The trip was coming to an end, and we were going back to Lautoka the next day.

Setariki wanted to leave early because he wanted to make a slight detour. He asked whether it was OK for us to take a bag of copra on behalf of the chief. Yeah sure, no problem. There was a small jetty on the island, and high tide right then allowed just enough water to get 'Dorado' alongside. It was a big bag, and it took three big guys to get it loaded into the cockpit. Well, it must have easily been 100kg.

Setariki's detour was to go fishing. He had a special spot he wanted to visit. It was a special spot all right: within an hour he had landed five big fish, must have been 2kgs each. He was going to sell them in Lautoka.

We arrived in Lautoka in the late afternoon and unloaded the cargo.

There seemed to be a utility truck nearby… Was that a coincidence?

There were no mobile phones in those days. I was quite impressed with Setariki. I'd been told that he was a boxer and looking at his physique and obvious fitness I could believe that he would be fairly handy, but he was also a very diplomatic guy, which he needed to be as security officer on board the Blue Lagoon vessels. 'We'll be up the pub later,' he called out as they left. 'Come and join us.'

Left on my own for the first time in 10 days, it was time to clean up, and start planning. I'd done everything that I wanted to do in Fiji. It was time to get ready and go. I looked round the boat carefully. Everything important seemed to be there. I knew that there had been people on board while I wasn't there. I'd gone back to the boat one night and found a group of them in the cockpit huddled around the compass. They'd switched the dark red fluorescent compass light on and were transfixed as the card danced around with the slight movement of the boat swinging to an anchor. I had some traveller's cheques squirreled away, and they were OK. Dave Mitchell had left a t-shirt lying around, and that seemed to have disappeared. Maybe that was the only casualty.

So, this was the plan: Go and see Customs in the morning… I had enough stores on board; just get some fresh bread and milk, and we'd be on our way. So, what about tonight? I'd noticed the Chinese restaurant just along the way; maybe they'd have take-away, or if they're not too busy I could just sit down, just for a treat on my last night. I got myself as presentable as possible, and off I went. The restaurant didn't look at all busy, and when I enquired, they were happy to serve me. I had a very nice meal with a couple of beers to go and was starting to feel good. Should I go up the pub? Yeah, why not, just a couple; thank Setariki for showing me around.

It's not far to the pub, and I'm soon walking in the door. It was quite busy at the bar, and I can't see Setariki anywhere. Oh, there they are, a group of them sitting at a table. 'Over here, George, we've kept you a seat.' I sat down between a Fijian guy on my right, and a young Fijian lady on my left, probably about 20-21, and quite good looking, I thought. 'Hello George,' she says. 'My name is Leina. I'm looking after you for the evening.' Oh

Chapter 3 - Dorado

really, I thought. this could be interesting. Somebody got me a glass and filled it up from a big jug of beer. Leina started talking about Sydney. She had visited Sydney, apparently, and wanted to go back. She knew people there, she said. They lived in Marrickville. I was remembering what Alf had told me before we left Sydney: Under no circumstances bring any of them back to Sydney. They will cause you no end of trouble, and the authorities will hold you responsible. Don't do it! Well, we hadn't reached that stage yet.

All of a sudden it was closing time at the pub, and everybody was going. We're going to Wilsons, somebody said. 'You're coming with us.' I was remembering what Alf had said about Wilsons, but I'd had several beers by now, and I had this young lady hanging onto my arm. 'Take me to Wilsons please, George,' she said. That was it then, the die was cast.

It wasn't hard to find the place; you could hear the music blaring out streets away. Up three flights of stairs, pay the entry fee and in. Strobe lights flashing, music booming, people dancing. Leina was trying to say something, but I couldn't hear what she was saying. She wanted a drink. OK. I was the only white face in the place, and I was copping a lot of glances, they weren't all friendly. I hadn't seen Setariki anywhere. He was here somewhere, presumably.

I don't know how long we'd been there… the noise was overpowering, and the continually flashing lights… There was a commotion, then it was on for young and old. I tried to get Leina away from it. Setariki appeared from nowhere and herded us into a corner. It was pretty wild out on the dance floor. People were disappearing over the balcony, three floors up. I found out later that they were dropping down into the Chinese restaurant, then coming back up the stairs, and re-joining the fray. The bouncer had no hope, he was just overrun.

Eventually, it settled down a bit and Setariki got us down the stairs and out on the street. Sorry about that, he said, it was totally out of control. I told him that was OK. 'It wasn't your fault,' I said. 'Thanks very much for getting us out safely'.

'George,' he says, 'I won't be seeing you again, so I hope you have enjoyed the last couple of weeks.'

'I certainly have,' I told him. It had been a great experience, and I thanked for that, too. A wave of the hand and he was gone. That left me there with Leina hanging onto my arm. I was about to say we'd better move away from here when she grabbed my arm even tighter and said, we're going back to my place. It was a fair distance back to Leina's place on the edge of town, but the walk was worth it. The dawn was up when she finally fell asleep, and I managed to slip away.

It took me ages to walk back to the wharf, and when I got there, I discovered that I wasn't alone anymore. Two Kiwi yachts had arrived. One of them was right alongside me, and I said good morning as I clambered aboard. I just about collapsed into my bunk.

It was 2pm when I woke up. I'd meant to leave that morning but there was still time to do it. I grabbed the various papers and headed over to the Customs House. It was close by, just next to the Blue Lagoon Cruises. That didn't take long, then it was around the corner to a general store to get some fresh bread and eggs. That was going to have to do. I got back on board and stowed the stuff away, then back on deck. No time for sail preparation right now, just take in the mooring lines and fenders, and off we go.

It was about 3:30pm, and we still had more than two hours of daylight left. I headed down the coast to where I had anchored on my way up. A nice safe anchorage for the night, get myself sorted out, have a good night's sleep, and be on my way in the morning. Slightly illegal, but who was going to know. Usually, I have several days beforehand to psyche myself up before leaving on a voyage, but the last few days had been quite stressful, and I wasn't ready mentally for this, my most significant voyage so far.

After good night's sleep, I was up early organising things. Charts and navigation first. The course to Sydney was fairly close to sou'-west, which meant we would be on a beam reach while

Chapter 3 - Dorado

we were in the trades, at least to 23 deg S. As far as sails were concerned, I was expecting to have fresh/strong trade winds, meaning I would need a reef in the mainsail, and the #3 jib. There would be a bit of sea running, and 'Dorado' might need some work to control. I was expecting that the selfsteering gear might need some help.

I was glad to be getting away from Fiji. Things were definitely deteriorating there at the end. There was no wind where we were at anchor, but I knew the wind would be there as soon as we passed through the reef. The anchor came up without difficulty and I stowed it away. It wouldn't be needed again. We had about an hour's run to the passage through the reef, and I used that to get the sails ready to hoist. Through the passage, and straight away the breeze is there: 15-20 knots sou'-east. Stop the engine and hoist the main. Raise the jib and set. 'Dorado' is moving fast. There is a bit of sea running. 'Dorado' is surging ahead with the swell. Now for the self-steering gear... set it with the vane feathering into the wind. It's working up to a point, but I have to keep adjusting things. I can't leave it alone for very long.

As we move away from the coast, the breeze increases to 20-25 knots. I put the second reef into the main. The mad rush eases a bit, and she's more manageable. The trouble is I can't leave the cockpit for any length of time. The adjustments I can make are 1) Adjust the vane on the selfsteering gear; 2) Adjust the mainsheet or take a reef in if necessary. This is the major adjustment; 'Dorado' is very sensitive to the mainsail, and it needs continual adjustment to get the best out of her. (If we were racing with a crew there would be a hand on the mainsheet at all times, but I'm singlehanded, and I have to compromise. I have to ease the sheet or put a reef in earlier than if there was a crew. This affects performance, which is something I have to accept); 3) I can set the tiller up with a small amount of weather helm to ease the strain on the self-steering gear, but it is dangerous to leave the tiller unattended in that position because a slight change in the wind force or direction could cause a wild jibe causing broken gear, or even the mast in extreme circumstances; 4) Adjust or change the headsail.

When I bought 'Dorado', she had a mainsail with roller furling, a genoa, #2, and #3 jibs, a spitfire jib, and a storm trysail, and not forgetting the spinnaker, which did get some use. During the years that I had her, I added a second mainsail, a 'yankee' jib, and a staysail. 'Dorado' had an inner forestay and could be rigged as a cutter. This gave me plenty of options for various wind and sea conditions. Generally speaking, the fore sails don't affect the steering of the boat very much. It's a rule of thumb that if you are reducing sail, you should reduce both fore and aft to keep the boat in balance.

I made a big effort in the first few days to rig the self-steering to control the boat, and to maintain a course, but it just wasn't powerful enough. The fresh-strong trade winds with a bit of sea running were just a bit much for it. I wanted something that could generate more power than just a wind vane, and I knew what that was. I had no option now; I was going to have to get an Aries from the UK. Anyway, we were still a long way from Sydney, and there would be plenty of time for further experimentation, especially when we hit the westerlies, closer to the Australian coast.

In the meantime, I wasn't getting enough sleep. There were times when I could rest and take it easy for a while, but still be in easy reach of the tiller and the mainsheet, and keep an eye on the compass, but going to bed and sleeping was out of the question. How long could I keep going like this? Not forever, that was for sure, but we were making good progress, and I didn't want to pass the opportunity up. I kept going for another couple of days, but exhaustion was setting in. What to do? We weren't that far from New Caledonia, maybe stop in Noumea for a few days and rest up. The opportunity is there now, let's do it.

I went to the charts and started laying a course. There was going to be an island (Walpole Island) in the way. I'd have to round it to the north, then due west would take me south of the reefs that extend well south of the New Caledonian mainland, then around to the Amédée Pass. There was another option: to go

Chapter 3 - Dorado

through a pass at the south-east corner of the mainland, but it would mean running straight at the pass in this fresh to strong wind. You'd get one go at it, and last year's events were still fresh in my mind.

We came up on Walpole Island during the night. It's uninhabited, so there were no lights, navigational or otherwise. The sky was clear, and there was a full moon. I saw it quite clearly and eased sheets a little and passed to the north. No problem. Now it's 120nm to the southern end of the New Caledonia reef. Paying particular attention to navigation, we're going to be passing the danger point during the night. About 3am, I can hear the boom of breakers. I harden up the sheets a bit and adjust our course 30 degrees to the south. The noise of the breakers fades slowly. Keep going for another hour… That's enough. Reset the course to north. The breeze is easing; we need more sail. I take the reef out of the main and hoist the genoa. Now we're doing a bit better.

The dawn was coming up and I could see land to the north, and waves breaking on the reef. Adjust the course to north-west and sail parallel to the reef. A friendly fishing boat comes by. Noumea? La bas! La bas! An arm waves to the north. The breeze is fading right away. Take the genoa down and start the engine. Leave the main up—it offers stability in confused seas. Another hour and the Amédée light comes into view. It is about 12nm from there to Noumea, so it's into the afternoon before I arrive at the yacht club. I have to wait for Maigret to come, and by the time that's done it is too late for anything else. Just sleep and rest and get up fresh in the morning.

I feel much better in the morning, and spend some time getting ready, then I go over to the club and see the lady in charge. As a visitor I can stay for a week for free, that should be enough time for me to rest up.

The last 10 days or so had been stressful, and there was a reaction to it. I was starting to think I couldn't go on. I'd leave 'Dorado' here and fly back to Sydney… come back and get her later. What kind of talk was this?

I had to overcome this depression. Just a few days' rest, taking it easy, A strong front came through Noumea; it would be blowing a gale out to sea. Give it a couple of days to settle down. I've got two free days left at the yacht club. I'm getting everything ready today, including some fresh food. All the gear is checked over, ready to go. Round to the Customs first thing in the morning; formalities completed.

The breeze is still pretty fresh. Will it be favourable? I'm hoping to make it to Amédée on the port tack. Tacking among the coral heads is not fun. There is always tension at the start of a trip, maybe more so this time.

Cast off the mooring lines, head out around Ilot Brun. The course from Ilot Brun to Amédée is almost exactly due south, and the breeze is sou'sou'-east at 15-25 knots. 'Dorado' can get to within 40 degrees to the wind, but that's not good enough, and we need four tacks to get down there. It's done, and we're through the pass, and the sheets are eased, and we're off and running at a good lick. Full mainsail and #2 jib.

There is not a lot of sea running, and the self-steering gear is doing better than earlier. The breeze eases slowly, and by next morning it's from the east at 10 knots. I raise the spinnaker (memories of last year) and continue on our way… the self-steering gear is useless in these light conditions.

Late in the day, the breeze shifts to nor'-east 10 knots. I hold onto the spinnaker for now. I've been listening to the marine weather forecast.

There is a strong Southern Ocean cold front sweeping the southern states. It will affect us up here in the next couple of days or so. The breeze slowly increases to 15 knots overnight. I'm hanging on to the spinnaker, but it's getting a bit interesting. The breeze shifts to the north during the morning; still hanging on; the westerlies are coming, won't be long now. Late in the afternoon the breeze shifts to the north-west at 20knots. Get the spinnaker in quickly and hoist the #2 jib and put the first reef in the main. It's been a great run for three days, and we're still going great guns. Still on course for Sydney Heads.

Chapter 3 - Dorado

The ocean forecasts are talking about gales in southern waters, but we are a long way north, and we should escape the worst of it. We hold the nor'-west wind for another couple of days and make good progress. More than halfway there. Navigation has been good; sights have been accurate.

That's the end of the joyride, and this is where the story really starts. The wind has shifted to the west at 25-30 knots, and the sea is really starting to build up. We're still laying the course for Sydney Heads, but we are now close--hauled, bashing into the steep short seas, spray flying everywhere. I change sails when the wind starts to freshen. Now we have two reefs in the main, and the #3 jib. This is the time of year for the westerlies, so I might as well get used to it. I was like a zombie by the morning. This couldn't continue.

The self-steering gear was no good in this situation. I was going to have to heave to for a while and get some rest. (To 'heave to' is to stop by backing the jib against the main. Tying off the tiller, the boat will sit relatively quietly in the water, until you are ready to move on.) I got about three hours sleep, then got up and had some food with coffee to go. Then I got the sextant out and did the morning shot and brought the navigation up to date. Feeling much better, I got under way again, until it was time for the noon latitude reading. It was best just to heave to for 10 minutes or so and get this important sight done and grab a coffee and a bikkie, then carry on until it was time for the afternoon shot and get the navigation up to date.

This was the routine for the next few days. Progress wasn't great; it was just about hanging in there and toughing it out. A good self-steering gear would have made a huge difference, and I knew that if I was going to be serious about circumnavigating, I was going to have to get one, indeed the best.

Eventually, the wind shifted to the sou'-west and started to ease. We were no longer able to lay the course for Sydney and were getting forced further offshore. The sensible thing to do was to get onto the port tack, and at least get a bit closer to the coast. Listening to the ocean forecast, the cold front had passed on and a large zone of high pressure was building across Australia. This would mean that our sou'-west wind would shift to the south then sou'-east over the next few days. That was good news, so I got up and shook a reef out of the main and replaced the #3 jib with the #2.

Progress improved, and in two days we were approaching Port Stephens. The breeze was now quite light, and even with help from the current it took us two days to reach Sydney. There was cloud and rain showers around so there was no celestial navigation, but I had seen the Norah Head light during the night, and I was confident we were somewhere off Sydney's northern beaches. There was one particularly heavy shower which cleared leaving me a good look at where we were: just past Long Reef, and North Head straight ahead.

The breeze, which had almost faded out, then picked up and gave us a very pleasant sail into the harbour. It had taken us 15 days to come from Noumea, which is a slow time, but understandable in the circumstances. I was quite pleased with the way I'd hung in there when the going got tough. I was back in business.

One of the great things about the cab game is that you're never out of a job. Two days after I got back to Balmain, I was driving again. Didn't I need a rest? After the last few weeks, driving a cab was like having a rest. This was now June 1976, and I was thinking that I could start my circumnavigation by April/May next year. I had to build a bank, and I reckoned that I could be away for three years, and I needed at least $15,000. I might be able to find work along the way, but that couldn't be guaranteed. I intended to leave Sydney with the boat in tip top condition, and with stores that would last for at least six months.

One of the things against me was that the good times in the cab game were gone, and we were in the middle of a worldwide recession. Anyway, go and get stuck in. What else can you do?

I'd already been in touch with Nick Franklin, the supplier in the UK, re the Aries self-steering gear. He seemed to be a very

Chapter 3 - Dorado

nice guy. He was going to send me an order form which I would complete and return with a cheque for $A3,500, which would cover shipping. I could expect it to arrive in Sydney in 8-12 weeks. I also started to plan and buy the charts that I would need, which was a quite a considerable number, also the sailing directions for the various ports that I would visit.

I was planning a 'classic' circumnavigation, taking the trade wind route. This has been done thousands of times, and many books have been written, and I've read many of them, but the ones I liked best were by a British couple, Eric and Susan Hiscock, who spent a large part of their lives sailing round the world in four different yachts all named Wanderer. Their books were full of information and advice about all aspects of cruising, and my voyage was planned based on their circumnavigation in Wanderer III, a 26ft timber yacht.

I'd sent off the order form and cheque to Nick Franklin, and in return he'd sent me a receipt and information about shipping, etc. It seemed that the ship carrying the Aries would be arriving in September, but of course the ship had to be unloaded, and the containers had to be unpacked and receive customs clearance. So, it was October before I got a note from the customs telling me that my goods were ready for pick up at their Villawood depot.

It was a Friday afternoon, and I was due to start work at 3pm, but my gear was there, and I wanted it now. I got in my car and drove out to Villawood, found the Customs depot and parked the car. I went into the building. I was directed to an office. The guy there sat me down and we went through some formalities, then he started on a long story about how my gear came under a special category, and there was a custom loading of 30 per cent... blah, blah, blah. I'm thinking, this bastard is just bullshitting to me; all he really wants is a sling. I was trying to find a way of putting it to him, but we were getting nowhere. Eventually he stood up and said, 'Ah just a minute,' and walked out the door. A few seconds went by, and another guy came in and sat down. Mr Brown, he says. Its Friday afternoon, and we're all looking forward to getting out of here and enjoying the weekend. This problem that you have can be resolved. We can get around it. It just needs some paperwork to be done. Now if you jump into your car and go up to that pub you passed on the way in and get two slabs of 'cold gold' stubbies, we'll have the paperwork done by the time you get back. That was putting it straight! I did exactly as he said, and when I came back, they needed my signature on a piece of paper, and then helped me get the gear in the back of the wagon. It had been a tough afternoon, and by the time I got back to Balmain, and unloaded the gear in a safe place, I was already an hour late for work.

When I'd sent the order form to Nick Franklin, I'd included a photo of 'Dorado's' stern and several measurements that Nick had asked for, so when I opened the box, and started looking inside I found the mounting pipes already bent to shape and ready to mount. That made the job of mounting it much easier, and it wasn't long before there it was, sitting on the stern of the boat, almost ready to go.

It looked big and bulky and heavy, but it was all made of aluminium and surprisingly light. The self-steering gears that I'd been experimenting with relied on a wind vane linked directly to the tiller.

The ones I'd tried were nowhere near powerful enough to overcome 'Dorado's' weather helm. The Aries had a wind vane which is mechanically linked to a paddle which hangs vertically in the water. The paddle can swing freely from side to side, up to about 60 degrees. If the breeze causes the wind vane to move the paddle will turn and its aerofoil shape will cause the paddle to swing left or right as the boat moves through the water. This swing produces a powerful force, and we harness this force, and connect it to the tiller, and use it to control the boat.

The Aries was the answer to my problem: controlling 'Dorado' at all times whilst we're under sail. It was the best investment I ever made. A robust piece of gear, it required little maintenance, and was still in good working order when we came to grief on Cape Horn.

Having got the Aries installed, the next move was to try it

Chapter 3 - Dorado

out. The harbour wasn't the ideal place because it was restricted, and even on a quiet day there was a lot more traffic. To get the best out of it, I had to go offshore, and offshore we went. Going offshore means a full day out from early in the morning till nearly dark. I did this a number of times (possibly as many as 10 times) in variable conditions, and I was very happy with the results.

The first time I went, it was calm when we left Balmain, but I knew there would be a nor'-east sea breeze later, and sure enough by the time we reached Bradley's Head, the nor'-east breeze was picking up. The sails were up and trimmed, the self-steering gear was set up and was working well, and we passed through the heads on the port tack. We held onto the port tack until we were well clear of the heads and all the associated chop, and everything settled down, then changed to the starboard tack. No problems: the Aries was in control. The breeze had steadied at nor'-east at 15 knots, a classic Sydney seabreeze. Next, I tacked back to port then eased the sheets slightly until we were on a beam reach. The seas were only slight, but 'Dorado' danced over them as we sped along at a rate of knots, and all under control of the Aries.

I was really enjoying this, but the afternoon was getting on, and we were passing the heads heading south. Tack again to starboard and head back through the heads on a shy reach. A very successful day and I'm very happy as we trickle down the harbour past the 'Sow and Pigs' (a small reef off to port as you come past Watsons Bay). Once past Bradleys Head, the breeze drops, and I get the auxiliary going and get back to Balmain as it's getting dark.

On another occasion, we went out on a day after a big southerly had come through, the wind had shifted to the sou'-sou'-east and was still blowing at 25-30 knots. I had one reef in the main and the #3 jib. The seas between the heads were very rough and lumpy, but they eased a bit after we passed through, although there was a decent swell to contend with. The Aries performed well on all points of sailing. I spent three hours out there close-hauled and reaching and running, and it was in control at all times. It was an impressive performance in the rough conditions, and I was very happy as we raced back up the harbour.

On one other occasion, we went out when it was rough. There had been a period of westerly weather, and the wind had finally settled from the sou'-west at 25-35 knots. the forecast said. Before we got to the heads, I had the second reef in the main, and the #3 jib set. The passage through the heads wasn't too bad, and we set sail down the coast towards Bondi.

We were being shielded by the high cliffs, and we were due for a shock. As we reached Ben Buckler, and Bondi Beach started to open up, the wind increased with a gust that must have reached at least 40 knots. 'Dorado' heeled over, and I was hanging on for dear life. The gust eased a little and she righted herself and continued on course. The Aries was in control and remained in control while I was struggling to stay on board. This was impressive.

The sea was rough, but the swell was subdued. Even so, it took us well over an hour to get back to the heads. We entered the heads and found the sou'-west wind blasting down that stretch of the harbour to Bradleys Head. We took it on and went tack on tack with the Aries in control. The Aries did a great job, and it wasn't long before we were rounding Bradleys and easing the sheets, giving us a shy reach all the way back to Balmain. A very successful day.

One thing we hadn't tried so far was to test the Aries out on a square run with the spinnaker up. I'd been waiting for a day of light winds and calm seas. Eventually we got a spell of calm weather, and my day off had a forecast of a light sea breeze 5-15 knots and calm seas. We left Balmain early and headed down the harbour hoping for the breeze to come in, but we had to motor all the way out through the heads before we got the first breath.

My plan was to tack upwind until we were 4-5nm off Long Reef, then set the spinnaker for a run back to the harbour. I had full main, and genoa set, and the breeze was very light from the east at 5-10 knots, so we started off on the starboard tack. I was

Chapter 3 - Dorado

expecting that the breeze would shift to the nor'-east and freshen a little, as usually happens. So, we're sailing along nicely. We're going to make it to Long Reef on this tack. Maybe the breeze will shift to the nor'-east by then. A very pleasant day.

The sea is as calm as you ever see it, just a very slight leftover sou'-east scend. Hardly noticeable. I get out of the cockpit and move forward to the mast. I'm looking at the leach of the genoa. It is fluttering slightly, as it should, according to the experts, if it's properly set. I'm standing there, not holding on to anything, leaning slightly backwards to compensate for the heel of the boat. I'm engrossed in the genoa, not paying attention to other things. There's a slight movement of the boat; I stumble backwards; a wild grab at the shrouds; I'm over the rail; I'm in the water; the boat is sailing past me; another wild grab misses, the stern is nearly past me; one more grab… last chance.

I grabbed the stern rail and pulled myself on board. I got into the cockpit and sat there for a while shivering and shaking. What the f--- happened then, feller! You were gone, 'Gone to Gowings'… What do you think you're playing at? It took me a little while to settle down, then I noticed that 'Dorado' was still sailing along straight towards Long Reef, and I had to do something about that. I had to get on with the day's plan, no matter what. I had to get back to Balmain… So, we'll do what I planned to do. We tack from starboard to port tack, and head out to sea, and away from Long Reef.

The breeze has freshened up to 12 knots east-nor'-east. We're moving nicely 5 knots–plus. We'll give it an hour or so, and then get the spinnaker up. On a racing boat, there would a hand on the halyard, a hand on the pole, two hands on the sheets, and of course a hand on the mainsheet. For me to do it on my own I had to have a system.

First of all, I would never attempt it in anything other than light weather, and calm seas. Get the boat set on its new course, ease the main right out, and secure it in case of an accidental jibe. Drop the genoa, but leave it attached to the stay (may be needed later). Get the spinnaker out of its bag and set it up in the bow ready for launch. Attach the sheets (port and starboard) making sure the loose ends are secured in case of accidents. Set up the pole at the mast and at the tack. Attach the halyard to the head of the spinnaker and haul it up. If you've done it all correctly, the sail should fill and start pulling you along. Dash back to the cockpit and adjust the sheets until it is setting properly.

The Aries plays an important part in this by keeping you on a steady course while you get the sail up and pulling. On this occasion, we discovered that, to lay the heads, we would be on a broad reach rather than a square run. 'Dorado' seemed to like that, and so did the Aries. We had a wonderful run back to North Head. As we were passing North Head, and South Head was opening up, I could see that if I changed course, we could run square with the breeze and go through the heads, then after passing South Head we could change back to our previous course and go right down towards the Watsons Bay/Vaucluse shore and across Rose Bay towards Point Piper. That would be a good place to drop the spinnaker and reset the Genoa.

As you approach the Harbour Bridge, the wind can be fickle, and the traffic concentrates. It's not the place to be fooling around with spinnakers. It was a wonderful day's sailing, but it was heavily qualified by my careless actions. I was happy to get back to Balmain.

I was pretty quiet for a while and didn't do any sailing. I started thinking seriously about safety, and what I should be doing about it. This was the mid-70s, and regulations around safety were not as strict as they were to become. I'd just paid lip service to things like safety harnesses, and life rafts, also radios. Bearing in mind that I was about to do a circumnavigation, I thought it was about time I updated myself. Also, the auxiliary motor, an 8hp Yanmar, wasn't very powerful, and we struggled to reach four knots in calm water. I was looking round to see what I could get with a 20hp output. I was talking to Alf about it one day, and he said, 'Funny you say that… There's a Yanmar dealer out at Rydalmere advertising two 18hp diesels, superseded stock.

Chapter 3 - Dorado

If you're quick, you'll get a good price. And I know someone who can install it for you'.

While all this was going on, a telegram arrived one day from my brother, Andrew. It read, 'Mum is seriously ill. She had a stroke and is in hospital more dead than alive. Please ring me on this number'. In 1976, there were no emails and no mobile phones, and I was going to have to walk up the street and use a public phone (if you could find one that worked). Clare offered their phone. I thanked her and offered to refund her for the call. The family wanted me to come, and I felt I should go, and so I went.

It was the right thing to do. It took a few days to arrange tickets and so forth. It was a return ticket, and I was due to come back in about a month's time. I wanted it to be two -three weeks, but that was the best I could do. It was 12 years since I'd seen Mum, and it was sad to see her in that condition. She was having difficulty talking, and her left side had been affected. She needed a calliper on her left leg. She came out of hospital about 10 days after I arrived, and it was obvious that recuperation would be a slow process. My father was there for her of course, but he wasn't in great health either.

It was good to see everybody, but it was wintertime, of course, and as soon as she was safely out of danger, my thoughts were turning back to Sydney. My plans to sail out of Sydney in April/May of this year were put on hold. I'd been struggling anyway, before mum got sick. Better to put it off for a year, and get everything done properly, and a decent bank to go.

Chapter 4

Circumnavigation

The extra year had gone quickly. I had worked six days a week to make sure I had an adequate 'bank', enough to last for three years or more if necessary. I hadn't done a lot of sailing, just Sundays, and a couple of trips up to Pittwater. I was very happy with 'Dorado'. She had a new auxiliary, a life raft of the valise type, which could be stored below, new wet weather gear with a harness sewn in, and other safety gear. Two new sails—a big yankee working jib, and a staysail to go with it. A new kerosene stove with plenty of spares to go with it. The old stove was gas, and the bottle that went with it needed a special connection to fill it. I couldn't get it filled in Fiji or in New Caledonia, and I ran out of gas halfway across the Tasman Sea. Not good! I'd even put a bowsprit on the bow as an experiment to see if it would help with the weather helm. I had all the charts I could reasonably require. Books such as those by the Hiscock's include lots of diagrams showing anchorages etc, and complement the official charts, and of course I had Alan Lucas's book, 'Cruising the Coral Coast' to see me all the way up to Thursday Island.

I laid in plenty of basic stores, enough to get me to South Africa at least. Everywhere I stop, I will get fresh bread, eggs, fruit, etc, but these don't last long when you go to sea. I get the compass checked. The steel bowsprit has made a difference, and it needs to be re-adjusted. I ring Captain Bozier's office. He doesn't come himself but sends someone else to do the job. No problem: the job gets done and paid for, and everybody's happy. It's early May 1978. The cyclone season is over; we'd be unlucky to get one now. The weather has been very calm and stable for some time, but there is a big cold front down in the (Great

Chapter 4 - Circumnavigation

Australian) Bight. Our local forecast here is for northerlies tomorrow going nor'-west, west, sou'-west over the next few days. That's exactly the weather I need to get me up the coast to Mooloolaba. I'm going tomorrow.

George Howie is in Perth. I give him a ring. 'Good luck,' he says. Others are coming down to the wharf tomorrow. A few beers in the 'Star Hotel' with Frank tonight. We haven't been so close since the 'John Vago' affair, but he wishes me well. I'm quite uptight tonight. I want to go and see Hilda, but her family are all there. She's done that deliberately. I phoned my brother in Scotland tonight. It's morning there. He'll let my mother know. That's the best way, I think. I hardly got any sleep.

In the morning, I walk up the street and have my usual breakfast at the usual deli. Delicious fresh rolls and salad, with coffee to go. I carry it back to Elkington Park, the usual spot. I can look down on the marina from here. I feel a breath of wind on my face. The northerly is in; time to go. I walk back down the marina. There are people there already waiting for me. Andy Brough is there with a couple of guys from Summer Hill; Joe Grech is there with Siriporn; Frank is there, of course, with Jack Buxton from Cameron's marina. Sue Brooks still lives on the houseboat; she's at the door, just a little wave. Start the auxiliary, release the mooring lines, manoeuvre out of the marina, we're on our way… they're all waving, and there's Hilda on her own at the end of Cameron's marina. Good on ya, Hilda, you've been a good friend to me this last year and a bit. You won't be at Balmain when I come back. Good luck to you too. Down past Long Nose Point, the breeze is in… get the main up. Under the bridge and past Neutral Bay, wave to the customs—not this time gentlemen; I won't be leaving Australian waters for a while yet, probably not till Thursday Island. Round Bradley's Head, enough breeze now to set the genoa (an auspicious start, full main and genoa), out the heads and harden up, port tack, course nor'-east.

The current will be against us nearly all the way to Mooloolaba, and it will probably take us 7-8 days. On the coast, there are the shipping lanes, but most danger comes from the fishing boats. Their movements are unpredictable. They have right of way if they are trawling, but not at other times. Night-time is worst; you just have to be on watch. The breeze holds north 10-15 knots; the sea is calm; we race along. In the morning, the breeze shifts to nor'-west 10-15 knots, and we're reaching now. Speed increases to 6-plus knots. That's fast for 'Dorado'. A couple of hours later, the wind increases to nor'-west 15-20 knots, and I drop the genoa and hoist the yankee and staysail. The sea is just starting to get up. We're past Port Stephens now, a fair way offshore, but that is OK.

By the evening of the second day, the wind is nor'-west 20-25 knots and the sea is starting to build. Progress is good, but the sea is short and rough. I put the first reef in the main before dark. The Aries is magnificent; it controls the boat at all times. I'm changing sails, trimming sails, taking sun sights, cooking meals, watching for traffic, having little rest periods. During the night, the wind shifts to west 30 knots. I have to put the second reef into the main during the night. The wind is gusting over 30 knots, the seas are rough, and spray is flying everywhere. There is not much likelihood of fishing boats being out in this weather.

The weather continues for another two days. On the fourth night, we pass South Solitary Island, off Coffs Harbour. The bright light gives us a definite fix. Another 24 hours and we pass Cape Byron. The breeze is starting to ease now and shift to the sou'-west. Shake a reef out of the main.

We're off the Gold Coast, and the breeze has left us

Chapter 4 - Circumnavigation

becalmed. Start up the auxiliary and make some progress. The sea is fairly calm and we're going along at five knots. A few hours later, we get a breath of wind from the sou'-east. It freshens up, and we're sailing again. Past North Stradbroke Island... We have to round Cape Moreton, and head nor'west to Mooloolaba. The breeze has come around to the nor'-east and is forcing us towards Moreton Island. We have to tack, and head offshore. We go for an hour, then tack back. The sky has gone cloudy, and the barometer is dropping. The wind has gone north and is starting to freshen.

It takes us another couple of hours to get around Cape Moreton, by which time the wind is nor'-west 25 knots. So, it's tack, tack, tack, all night. One reef in the main, and #3 jib. The seas are not too bad, there is no residual swell. At daylight the wind shifts to west 25 knots, and for the first time we can lay the course to Mooloolaba.

As the morning progresses, the wind starts to ease, and shifts to the sou'-west. We are closing with the land, and I can recognise features. We come around Point Cartwright, and the entrance to Mooloolaba harbour is obvious. Drop the sails, start the engine, line up the leads, and in we go. The yacht club is not far inside the harbour, tie up at the end of the wharf, and go see the people at the club.

'Yes, you are welcome,' they tell me. 'You can stay for a week for free.' There are several free berths at the moment; I take #26. We move into the berth and set up the mooring lines. Now I can relax for a while and have a cup of coffee and reflect. We did it in just over six days. That's very good for 'Dorado'. It was a bit rough at times, especially last night, but that's behind us now. 'Dorado' sailed very well, and the Aries performed magnificently.

The bowsprit was not a success. It was not affecting the

weather helm as far as I could see, and it was affecting the setting of the jibs. The foot of the jib was coming over the rail, and chafing, and in fact I was going to have to do repair work on the genoa, also #2 and #3 jibs. I was going to reset the forestay back to its original position. Should I take the bowsprit off, or just leave it there doing nothing?

Right next to the yacht club was the fisherman's co-op, and part of the co-op was a fish and chip restaurant. Beautiful food, absolutely fresh fish, beautifully cooked and presented. Just what the sailor needed after six days at sea. The yacht club had a restaurant, which was OK, and there was a pub not far away which sold hot pies. I didn't trouble the pub in the few days that I was in Mooloolaba.

I was finished my sail repairs and starting to get ready for my next move. I wanted to get to Cairns, but I wanted to break it into stages, with little rests between stages. In Alan Lucas's book, Scawfell Island is mentioned as a good, safe anchorage, and that seemed like a good start. That would put me in a good position to go right round the outside of the Whitsundays and head up towards the Townsville area. I'm not that fussed about going into Townsville; maybe we could anchor at Magnetic Island. The weather is fine, and the forecast is for light sou'-east winds, I'm going tomorrow morning.

Up bright and early, off to the shops for fresh bread and eggs, also fresh fruit. Scawfell should take us about four days, but there'll be no more fresh stuff until Magnetic Island, maybe 10 days away. Get the sails ready to hoist, engine started, let go the mooring lines, out of the berth. . No breeze at the moment. Our course is just about due north for the next 120nm, at least 24hrs

Chapter 4 - Circumnavigation

away. We are travelling up the Fraser Island coast all the way and past the Breaksea spit area. Then we head nor'-nor'-west for about 100nm which will bring us up to the east of Lady Elliot Island. There is a light on Lady Elliot Island which will be a great help if we approach after dark. From Lady Elliot Island, the course is nor'-west around 250nm, and that will take us all the way to Scawfell.

Round about lunchtime, a very light breath of sou'-east breeze came in, and in another hour, it had freshened up enough for me to hoist the genoa, stop the engine, and sail. We were on a reach, and 'Dorado' was sailing at her best. It was the next morning by the time we reached the Breaksea spit area, to which we gave a wide berth and continued on our new course of nor'-nor'-west. We should have passed Lady Elliot Island, during the night. I should have seen the light at least 10nm away, but I saw nothing. I was getting quite concerned. I didn't want to run into the reef, and I kept a very close watch for some hours. Finally, the dawn came up, and nothing in view. Keep going for now, but as soon as the sun gets higher, I'll start taking sights.

I started to suspect that the compass was out. In a last-minute decision in Mooloolaba, I took the bowsprit off. Why do I do these things? There was a good chance that my messing around with it has affected the compass. The log was showing well over 100nm since we left Fraser Island. When I got my sun sights, it was clear that we were well past

Lady Elliot Island, just a little wider than I expected to be. (I later discovered that the Lady Elliot light had not been operating for more than a year.) Having got ourselves sorted out, I changed course to nor'west and settled things down. There were no dangers now for over 150nm so I could focus on navigation and checking the compass. Weather conditions were still light to moderate, and we were sailing along fairly upright (compass errors were more likely to show up when the boat starts to heel). I got the hand bearing compass out and tried to check the yacht's compass against it. The results were inconclusive, and I decided that it was OK for now.

Sailing at night was amazing. The moon was nearly full, and the water was full of phosphorescence. The wake stretched out behind us, blazing with colour. We didn't see many ships, just a few fishing boats, and we had no problems with them. We were in the trade wind zone for sure, and the wind remained steady sou'-east 15 knots. We made good progress, and by late the next day we were passing the Percy group of islands, coming level with Sarina then Mackay on the North Queensland coast.

I now had to pay close attention to navigation as there were a lot of islands and reefs off this part of the coast, and many we would come close to before we reached Scawfell. Probably the most important will be Penrith Island, about 70km off the coast and level with Mackay. It's quite high and can be seen from a good distance. If we can line ourselves up with that, we pass it to starboard, and continue for another 10nm on the same course, and arrive at Scawfell.

We reach Scawfell in the late afternoon. I was disappointed to find there were several yachts already there. The best spots to anchor were taken, and I was forced to anchor farther out in much deeper water, where there were also coral heads to contend with. The water was very deep (more than 40', from memory), so I had to let out all the chain, and then some rope before I was satisfied. There were strong gusts of wind coming down off the land, but the anchor seemed to be holding OK. It had been three days without much sleep, and as soon as I had something to eat, I went to bed.

Chapter 4 - Circumnavigation

I woke at dawn with a vague feeling that something was wrong. I went straight up on deck, and immediately saw that we had moved. I went up to the bow to check the anchor. The rope was just hanging there. I lifted it up, and it came up, light as a feather. I lifted some more, and a frayed end came out of the water. It felt all slimy, maybe rubbing over coral, or perhaps chewed through by something. Reality was setting in. Right now, we were slowly drifting towards the rocks.

I got the engine going and backed off a bit, just to think it out. I shouldn't have used the rope for anchoring; it would have been better used as mooring rope. I didn't let out too much of it. I didn't think it would reach the coral. Maybe something chewed through it; is that possible? So, what am I going to do now? I have another anchor, a smaller one, with lighter chain. Should I anchor temporarily and try and retrieve my lost anchor? Could I dive down 40' or more, and what would I find there? I could see from the surface it was all dark down there, no clean sand anywhere. It was all too hard for me. I accepted the fact that the anchor was gone and got the sails up and sailed away. (Had I sailed into the Scawfell anchorage at the end of the circumnavigation, with the benefit of experience, I would have gone right up among the other anchored yachts, dropped anchor between them, and then sat back behind them as I drove the anchor deep in the sand. If I'd had to drop it farther back amid the coral, I would never have used rope, under any circumstances. And if it had gotten stuck for some reason or another, I definitely would have tried to retrieve it. I'm afraid I was still very much a learner at this cruising business!)

So, I was going to have to go to Townsville, whether I liked it or not. It was a trip of around 200nm. There was one island along the way which I could use as a navigation check, and that was Holbourne Island, and we should be sighting it the next morning. Our course was north from Scawfell until we passed the Whitsunday group, then nor'-west until we passed Holbourne to port, then about wes-nor'-west until we reached Cape Cleveland, and into Townsville. Alan Lucas says that pile berth moorings are available in Townsville, and that would be best for us since we've lost our main anchor.

The breeze was sou'-east 15-20 knots when we left Scawfell, and we made great progress until we were past Holbourne Island. From then on, the breeze started to fade, and we were left nearly becalmed between Cape Bowling Green and Cape Cleveland. I had to start the auxiliary to get us into Townsville before dark. I had no trouble finding the pile moorings and finding one that was available at reasonable cost. Rest was the order for the first couple of days, and I needed it badly.

Townsville is a sizeable place, with a population of around 100,000. There was a bit of industry there, a large military base with service industries, and a large busy harbour. People at the marina told me where I could get a new anchor and gave me directions. I walked out there, as there didn't appear to be any public transport. The place was well stocked, and had what I required, no problem. The previous anchor was 25lb, and the chain was 5/16". I had read Hiscock's books of course, and for a 35' boat he recommended a 35lb anchor and 3/8" chain. I got about 200' of chain, and again I was guided by Hiscock.

I would need transport back to the harbour and I asked the fellow in the shop about calling a taxi. 'Don't do that,' he says. 'My son is just about ready to go into town. He'll take you.' That was lucky, and he took me right down to the wharf. I just had to

Chapter 4 - Circumnavigation

get it into the dinghy and out to the boat.

By the time I got the stuff out to the boat and properly stowed away, I felt I'd done a fair day's work and deserved a drink or two. There was a bar ashore, and I'd noticed some yachties there earlier on. A few drinks, swap a few yarns, a little bit of company, makes all the difference. I slept like a log and woke up refreshed in the morning. Now the other thing I need to get done is have someone look at the compass. I'd met a guy yesterday, and he said, yes, there were compass adjusters in Townsville, and he wrote down a name and a phone number for me. I rang the number the next day and the fellow said, yes, he adjusts compasses. We arranged a place and time, and went out into the bay, and got the job done. The report said the compass wasn't too bad, but there were minor errors. The main thing was that the compass hadn't been set up very well the last time that it had been done, and he'd replaced the powerful vertical magnets that are set up beneath the compass. He also described to me a method of checking the compass using the sun. I'd read about this method in one of Hiscock's books, and used it, especially on ocean crossings, and in areas of high magnetic variation.

That few days in Townsville had cost quite a bit of money that I hadn't budgeted for, and it reminded me that I wasn't a millionaire and had a fixed budget. Time was up in Townsville, and it was time to move on. Next stop was going to be Cairns. It was about 156nm, and a fairly straight-forward trip. I left Townsville and went out to Magnetic Island for a couple of nights, then set off for Cairns early in the morning. We followed the shipping channel this time, which runs just about due north all the way up to Fitzroy Island, then follow the coast around till you pick up the dredged channel which takes you into Cairns' harbour (Trinity Inlet). We encountered a bit of shipping on this trip but had no problem with them. It's the fishing boats that cause the biggest problem. Anyway, we got there safely.

My memories of Cairns harbour are a bit vague. I seem to remember that I was allowed up at the main wharf for a few days. The Navy wasn't in Cairns in those days, and there was no shipping due for more than a week. The wharf was designed for ships, so even at high tide it was difficult to get off the boat. I was lucky that I was able to moor next to a ladder. I wasn't all that impressed with Cairns and stayed only for a few days.

Our next stop was to be Cooktown, an historical site going back to the days of Captain Cook. About 90nm north of Cairns, it was going to be another overnight trip and there was a danger spot at the Hope Islands, where the reef comes very close to the coast, and the shipping lane is very narrow, as little as a couple of kilometres in one place. Maybe we could go as far as the Low Isles and anchor for the night. Let's see how we go.

The route is a slight dogleg, nor'-west from Cairns to the Low Isles then nor'-nor'-west from there to Cooktown. We leave early form Cairns, negotiate the channel, and we're on our way. The breeze is sou'-east 20 knots and we move along nicely. We're approaching the Low Isles by late afternoon. Should we stop and anchor? Nah! She'll be right! It wasn't long after we passed the Low Isles that we met our first fishing boat, and we had to change course to avoid it, and then another, and another. They're all out tonight, and this is their favourite fishing ground. It was like dodgem cars for a while.

It was after dark now, and were approaching Gubbins Reef, that notorious place where the 'Endeavour' had run aground. I'm just lining up to pass through the narrows when I see the lights of an approaching ship. No way, I can't get through there before the ship arrives. We veer off to port… Where are we? How far can

Chapter 4 - Circumnavigation

we go? We can't be far off the reef. The ship is passing through. It'll be gone in a couple of minutes. Hooley Dooley! There's another ship coming! Another nervous wait for 15 minutes or so.

Finally, we get through that narrow stretch of water, and proceed towards Cooktown. The fishing boats are not finished with us yet and there are more nervous moments before we enter the river and drop anchor. There are other yachts around, and I anchor among them. Looking at the shore, it looks like about half tide (tidal range is bigger here than in Sydney). I should be OK here for a while.

Cooktown is the last outpost of civilisation before you get to Thursday Island. It's remote and feels remote. It has a charm of its own. I spent a restful week there walking around, soaking in the history, and the beer at the pub. The locals were friendlier than Cairns, or anywhere else that I'd been for that matter, and I enjoyed it. I'd been there for nearly a week when I realised, Yeah, this is great, but don't get sucked in. It's time to be moving on. I'd already decided that, after my recent experience, there would be no more night sailing on the east coast. The reef closes right in close to the coast, and it is bad enough during the day.

Our next move was to Lizard Island, which is about 70nm from Cooktown. There is a very good anchorage there, and there is more history attached to it, as the Endeavour stopped there. The trade wind had been pumping 25-30 knots for days, with no likelihood of easing any time soon, so we just had to get on with it. From Cooktown to Cape Bedford, just south-west of Lizard Island, our course was about eastnor'-east, and we were close-hauled, and in fact we had to make a tack about halfway along in fairly restricted waters, so we did it tough for that first part of the trip. After rounding Cape Bedford, our course was nor'-nor'-east, so we could ease sheets and race along.

As we approached Lizard Island we had to watch out for 'bombies' (bomboras, isolated, and unmarked coral heads, which lie just below the surface). I did see a couple, but passed them safely, and were soon at anchor in Watsons Bay, a beautiful and safe anchorage. Lizard Island is a special place, beautiful, remote, and peaceful. It's a quite large continental island, 360m high, and there are some good walks. I climbed to the high point and found a stone marking the spot where Captain Cook came with his telescope searching for a way out of the reef. There is also a small but exclusive resort on the island with an airstrip, catering for people who go fishing for blue marlin. I believe the actor Lee Marvin came to Lizard Island regularly over a period of years. There is also a small research station for scientists who study coral reefs.

From Lizard Island, we headed more or less west around Cape Melville and into Bathurst Bay, where we anchored behind Flinders Island. Again, this was a trip of about 70nm, which is do-able because there is a nice fresh trade wind sending you along. We were now about 250nm from Thursday Island, and this part of the voyage needed careful planning. This area is very remote, and great care is required. The reef closes with the coast and leaves little room for manoeuvre if there are other vessels around. Sailing at night was out of the question, but Alan Lucas has surveyed this area, and has detailed a number of anchorages behind coral reefs or cays, which are suitable for overnight stays. So, it was up early in the morning, and a dash of 50nm or so to the next stop behind a coral reef, and so on for the next few days, until we got to Adolphus Island, about 25nm east of Thursday Island.

Chapter 4 - Circumnavigation

I just paused at Adolphus for a day to take stock. It had been fairly hectic the last few days, and I just wanted to get myself ready for Thursday Island. Lucas actually recommends that you go to Horn Island and anchor. There is a wooden wharf, and a ferry runs over from Thursday Island two or three times a day, bringing passengers to the airport on Horn Island. There is good holding at Horn, and you are out of the wind. Sounded like common sense to me. So next morning we got away early and got over to the wharf at Horn Island and anchored among some other yachts.

Lucas was right: it was a nice, calm, secure anchorage. Next morning, the ferry came over with passengers going to the airport, and I got on board, making sure he was coming back in the afternoon. About 3pm, he told me. So, I had several hours to wander round Thursday Island. I found the Customs House and went in and introduced myself. I'd be coming to see them in a few days' time. There were a couple of grocery stores, and I could get fresh bread and eggs. There was a guy making fresh 'burgers' and I indulged in one of these. There was a place that showed movies, and guess what: there was a pub. I went in there and had a couple while I was waiting for the ferry.

Thursday Island is not the sort of place that you could fall in love with. It is a sort of ramshackle place where everybody is passing through on their way to somewhere else. Me too! Next day I stayed over at Horn Island and went for a walk out along the road that went to the airstrip. It was farther than I thought, and when I got there, there was nothing to see except a small shed that served as the terminal building. In the afternoon, I settled down with my charts and sailing directions and planned my next move, which I had decided was going to be to KeelingCocos, a large coral reef all the way across the top of Northern Australia in the Indian Ocean.

Meanwhile, back by Thursday Island, the planning was finished, and it was time to get moving. There was nothing to hold me in this place. I was going over to anchor at Thursday Island the next morning, and see Customs and Immigration, and get the hell out of it. It wasn't far but there were strong winds and currents, and shoals to contend with. The tide was most suitable around late morning, and I waited till then. There was no problem, and we reached the anchorage in about 20 minutes.

I was getting us lined up to approach the anchorage when something caught my eye. A small white yacht of about 24' was bobbing around on its anchor. That's a Clansman, I thought (a design very common in Sydney). What's it doing up here, I wondered. As I was passing it, the hatch opened, and a head came out. Shit, I thought, I know that guy. He was waving, and I waved back. I'll come over and see you, I yelled out as we passed. He waved back and shouted something that I couldn't make out. I found a suitable space and dropped the anchor. It was holding quite nicely, but I put the auxiliary in reverse for a while just to make sure it was properly dug in. Thursday Island has a reputation of not being a great anchorage; best not to take chances. I wasn't going to be here for very long, maybe just a few hours.

I could see the Clansman not far behind me, maybe 100m. I could get there alright; it would be coming back that would be tricky. I rowed over there in no time, and he was there to take the mooring line as I stepped on board. 'G'day,' I said. 'How are you?'

'I'm sorry, I don't even know your name.' (He'd been at Balmain for months, and Frank and I had tried to engage him in conversation to no avail). He introduced himself as Terry.

'Come below out of this wind,' he said.

Chapter 4 - Circumnavigation

We went below and started chatting, and the bottle of rum came out. He was English and had owned a much larger yacht at one time. He'd left England in that yacht and cruised to the West Indies, where he worked off and on for a few years. He was an electrician, but specialised in TV repairs, radio repairs, etc. Then there was an accident, and his yacht was burnt to the waterline. He was stranded in the West Indies but was lucky enough to be offered a position as crew on a French yacht heading to Tahiti. Tahiti is a bad place to get stranded, but again he was lucky and got a position as crew on an Australian yacht returning to Sydney. Some insurance money had come through allowing him to buy a cheap yacht in Sydney, and here he was in Thursday Island, on his way back to England via South Africa.

I suddenly realised that the rum bottle was half empty and I was outstaying my welcome. If I was going to leave Thursday Island today, I'd better get a move on. Are you going to Keeling-Cocos I asked him? No, he said, he was going to Christmas Island then the Seychelles. Ah, well, I said our paths might cross in South Africa. All the best. I got out on deck and went to get the dinghy, the dinghy... the dinghy was gone. I stared stupidly at the place where it had been. I'd watched him tie it onto the back rail, he used a clove hitch, the same knot I would have used, although I would have put in a couple of extra hitches for safety. The strong breeze was causing quite a bit of chop, and the dinghy must have tugged itself free. I looked up and down the beach—nothing. If it had gone around the point, it's gone for good.

The implications were coming home to me. I had to have a dinghy, and if I could find anything here it would be some astronomical price. The first thing to do was to get back on board 'Dorado'. Could you drop me off? I asked him. I haven't got an engine, he replied. I looked down the hatch to where the auxiliary should be… nothing. Surely, you must have an outboard. He shook his head. Could I borrow your dinghy then? He showed me his dinghy. It was tiny, and I'd go straight through it if I tried to get in it. I'm going to have to swim, I thought. It's not that far. There's another moored boat about halfway if I get tired.

Four or five shots of rum were giving me bravado that overrode common sense, I just wanted to get back to 'Dorado' and I jumped into the water and took off. I was going well and making some progress nearly up to the first boat when I heard an engine start up, then I saw a large launch circling around and coming up behind me. It pulled up alongside me, and a guy leaned over the side and spoke. 'Mate, have you got a f'ing death wish or something? Hang onto this line and I'll take you back to your boat.' It took only a few seconds, and I was clambering back on board 'Dorado'. Thanks very much for your help, I yelled out to him as he motored away.

Back on board 'Dorado', I just had to sit down for a minute and take stock of the situation. I was puffing and blowing fairly hard. I could swim, of course, but I wasn't a strong swimmer. Could I have made it all the way? Maybe, but we'll never know. I was going to have to get ashore, and that meant I was going to have to tie up at the main wharf. It wasn't occupied at the time, so I moved over there straight away. I hope nobody wants it for a while, I thought as I walked away.

I searched all along the beach but found nothing. I went right out to the point. Nothing there. Had it gone past the point, it would be miles out to sea by now. I wasn't going to find it; it was gone, that was obvious. So, the question now was, where am I going to get another one. I started walking the streets, to see if I could see any dinghies. Maybe I could buy one. What would it cost on Thursday Island? I hate to think. Quite a few houses had

Chapter 4 - Circumnavigation

dinghies in their yard; some were broken, some in disrepair, most were far too big. I spoke to a number of people. No-one could suggest anything. It was getting late in the afternoon. It would be dark soon. I'd better get back to the boat, try again tomorrow.

I was walking down this street past the police station... Yeah, maybe they might know something, just go in and ask. There's a young constable at the desk and I start to tell him my story. He doesn't say anything until I'm finished, then he says, 'Can you describe your dinghy?' I start to describe it, and as I talk, he walks over to a door behind him and opens it. There in the backyard, framed by the doorway, is my dinghy. I was flabbergasted, I couldn't believe what I was seeing. 'That's my dinghy,' I managed to say. The priest saw it coming ashore

on the beach, the policeman said, and brought it in. We knew that somebody would be looking for it. Ye Gods! I thought, how lucky can you be. I've got to go and see this guy and thank him. The policeman gave me directions. 'Pick the dinghy up on the way back,' he said.

It wasn't far to the chapel, and the priest was right there. I thanked him profusely and offered him some money. He wouldn't take the money for himself but showed me the 'poor box' and I put $40 in it. It was a lot of money for me at that time, but he deserved it. Then it was back to the police station. I got the dinghy on my back, the oars tucked under my arm, and back to the wharf. What a day that had been. It was just getting dark as we left the wharf and motored back to where we'd anchored previously. I'm doing no more today, a good night's rest, I'll go ashore tomorrow morning and see the Customs and Immigration, then some fresh food to take with me, and I'm off outa here.

'You'll have to show your passport when you get to Keeling-

Cocos,' the Customs fellow told me. 'Make sure you do. They've been cracking down recently.'

No problem. I was ready to go by midday. The trade wind was blowing strongly, and the currents were favourable, and we went shooting past the last of the islands at some amazing rate of knots. Clear of any danger, I got some sail up. The wind and the current were easing, and I set a full main and genoa. We passed Booby Island, a very well-known stopping place in the days of the sailing ships (there was even a 'post office' in a cave at the northern end). I get the binoculars out and have a look, yes, I see the cave; not stopping today though.

Overnight, the breeze fades away to less than 10 knots, and speed drops below four knots. Our course at this stage is about west-sou'-west, sea is calm. We're in the Gulf of Carpentaria, and I don't expect any large seas here. I take down the genoa and hoist the spinnaker. Speed improves, over five knots. We're reaching with the spinnaker up, OK, in light breezes and calm seas. The Aries is in control. I go below to update the log. I do this every hour (current reading/distance run/wind force/direction. Notes: spinnaker up, going well). Time for a bite of breakfast, and a cup of coffee.

The breeze is increasing. I can hear the log ticking away. I can feel 'Dorado' starting to strain. It's time I was on deck. 'Dorado' was racing along, speed over seven knots. She was also starting to heel over. We don't want the chine coming out of the water. That's when the drama starts. Ease the course a little to west. That's better; she's more upright now, still doing seven knots. The breeze held all day. As night approaches, I have to decide, should I take the spinnaker in? It's a safety thing. OK as long as I'm there to supervise; not so good if I go below and leave it unattended, especially if I go to sleep. I've seen no shipping in

Chapter 4 - Circumnavigation

the last 36 hours, no fishing boats, the sea is calm, the sky is clear, the moon is nearly full, the breeze has eased just a touch, we're still doing six knots. Leave the spinnaker for now anyway. We're making great progress.

Speaking of the moon, I'd promised myself that I would start using the moon for navigation, and now would be an excellent time to practise. The moon is an awkward thing to shoot because it is rarely full, and it is moving fast, and you need to allow for several corrections. These are all accounted for in the almanac, of course, and what counts is practise, practise, practise. It is worthwhile doing because the time might come when the moon is the only option available to help the navigator make a decision. Much the same applies to the planets, and the navigational stars, and I was to spend the next three weeks of this voyage becoming proficient and comfortable with all of these options.

Our great run lasted for another two days, by which time we had crossed the Gulf of Carpentaria and were north of Arnhem Land. The sou'-east breeze started to ease, and progress slowed. From about the time that we were due north of Darwin, we had variable weather for several days. Winds, mainly light/moderate from the westerly quarter, cloudy, thunderstorms around. Our course at this time was sou'-west. We needed to stay south of Timor, and also watch out for Ashmore Reef, which was a danger. It wouldn't be visible until we were quite close, and that was potentially dangerous.

I never saw Ashmore Reef, which is good. It's between Timor and the north of the Australian mainland, nor'-west of Darwin. What I did see were the high mountains of Timor, which told me that we were past Ashmore Reef and entering the Indian Ocean. The next day, the wind shifted to the south, very light to start, but slowly freshening and shifting to the sou'-east. We were in the Indian Ocean and the sou'-east trades had found us. Our course was now due west, and we were about 1700nm from Keeling-Cocos Islands.

It was two days later when a strange event happened. It was during the afternoon, and again I was below deck having a little rest. I became aware of a noise that was out of place. It was music, it was rock'n'roll, and it was getting louder. I shot up on deck quick smart to see what was going on. Off on the port side, about 300m away, there was a vessel that looked like an ancient fishing boat, and it was heading towards us. It was obviously on autopilot and was veering from side to side as it headed north. There was no sign of any crew. I watched for a few seconds and realised that it wasn't going to hit us. As it got near us the noise was quite overpowering, and as it passed astern by about 50m, there were the crew, about four of them sitting in the cockpit astern staring at me with glazed eyes, all high on drugs no doubt.

This was a potentially dangerous situation. There were at least four of them and could have overpowered me if they'd had a mind to. They were all off their heads on drugs and had nothing else on their minds, which was lucky for me. They would have come from the north-west coast of Australia where there are innumerable reefs and cays, where they find Trepang (as they call it). It is a great delicacy in Indonesian culture, and they have been doing it since time immemorial. In this day and age, it is technically illegal, but the Australian authorities turn a blind eye to it. Australian people refer to Trepang as 'sea slugs', which tells you what they think about it.

Chapter 4 - Circumnavigation

The days passed peacefully enough. The trade winds were a constant fresh sou'-east. Progress was good 140-150nm per day. I was concentrating on navigation, doing sun shots, moon shots, planets and stars, becoming familiar with the various almanacs and tables. About two weeks after entering the trades, we were nearing Keeling-Cocos and I reckoned that we might arrive the next day. I'd done my afternoon sun shots and I was below decks at the chart table plotting our position when suddenly there is a loud voice.

'Ahoy, Ahoy, Ahoy!'

I flew up on deck to find a huge motor cruiser circling around us. It was all white, and in black writing on the side was, 'Oil Industry Survey Vessel'. The vessel had three stories above the deck, and up on the top story was a guy in a white uniform with a captain's cap on. He had a megaphone in his hand and was yelling at me in an American accent.

'Are you OK?' he yelled.

Yeah, sure. Nod the head, thumbs up.

'Do you have VHF radio?' was next.

No, shake the head, no radio.

'Are you calling at Keeling-Cocos?' he yelled.

Yes, nod the head, thumbs up!

'If you're going to Keeling-Cocos,' he yelled, 'you need to change your course five degrees to south.'

I'd just finished doing the afternoon sun shot, and updated our position, and I knew we were headed straight for Home Island. You're playing to your audience aren't you, big mouth, I thought. (While he'd been yelling at me, a number of other people had appeared, and all were gawking at me). I wish I had a megaphone, I thought, I'd tell you lot where to get off. He went around one more time, so everybody could get a photo, and then headed off to the east. Another strange incident I didn't need.

Of course, I had to go and recheck my figures, and found them all OK. Nevertheless, the approach was going to be tricky. The highest things on the island were the palm trees, and they would be hidden by the heavy trade wind haze, until we were very close. According to my calculations, we would be arriving in the early morning, which was good. Arriving during the dark hours and trying to navigate inside the reef was out of the question.

I had a restless night, scared to fall asleep in case some drama arose. Getting on towards dawn, I saw some flashing lights some distance off, probably over the horizon. There is an airstrip on West Island, and planes sometimes land and take off at night. This is mentioned in the sailing directions and was just another indication that I was bang on course. It was about two hours after dawn that the palm trees appeared out of the haze, and I had to adjust course a little to the north. The pass through the reef is at the northern end of the group, around Direction Island. After getting through the pass, the coral starts to close in from all sides.

Cocos-Keeling Islands are a number of islands which had grown up on the reef. Vegetation, including coconut palms, had grown there over the years. A Scottish family, the Clunies-Ross family, had settled there in 1827 and claimed all the islands as theirs. They made a living out of coconuts and coconut products and brought in workers from Malaysia. So, a strange little society built up in this remote place over a period of 150 years. All good things must come to an end, however and, at the time that I was there, there were ongoing negotiations between the Clunies-Ross family and the Australian government which ended with

Chapter 4 - Circumnavigation

Australia taking over the island group.

The family lived on Home Island on the eastern side of the reef and were still there at the time of my visit. Over on the other side of the reef was West Island, which is fairly long and holds an airstrip. There are buildings there, which contain the usual functions of an airport, including customs and immigration. Cocos-Keeling was an important stop for me, breaking a long voyage across the Indian Ocean. I also intended to stop at Mauritius, before arriving in Durban in November.

I'd been told in Thursday Island that I would be required to show my passport, and I'd figured out the logical place to do that was at Home Island. With this in mind, I dropped the sails as soon as we were through the pass and started the auxiliary. There was no clear channel marked on the chart, so we proceeded very carefully towards a wharf that I could see on Home Island. There was coral beneath us all the way, and there didn't seem to be a lot of clearance. We were just starting to approach the wharf when we reached a large area where there was no coral, and a sandy bottom. This'll do us for the moment, I thought, and let the anchor go. It dug in nicely, and the holding appeared to be good.

I was just starting to untie the dinghy from its place on the cabin top when I saw a launch leaving the wharf and head in our direction. It came right over and pulled up alongside.

'I'm John Clunies-Ross,' the fellow in the launch says.

'You can't anchor here. Follow me and I'll take you over to the yacht anchorage at Direction Island.'

I hauled up the anchor and followed him as we wound our way through the maze of coral heads. As we got close to Direction Island, the coral gave way to clean sand.

'You can anchor anywhere here,' he called.

He waited until I'd anchored properly and then came over

alongside.

'Where have you come from,' he asked?

'Thursday Island,' I replied.

'They told me you'd be wanting to see my passport,' I said.

'Yes, you can give it to me now if you like,' he said. '

You'll get it back on the day that you leave at the passport office over on West Island, OK.'

He must have seen some doubt on my face and continued: take your boat over to the northern point of West Island and anchor there, then go ashore and walk to the airstrip where you will find the Passport Office. 'Enjoy your stay,' he called out as he left.

The anchorage at Direction Island was quite secure but was open to the trade wind. Direction Island itself was covered in palm trees and undergrowth, but the next section of reef going down to Home Island was completely bare, and the trade wind just howled across it. I'd been fairly lucky on my way over, the breeze being fresh, and sailing very enjoyable, but the day after I arrived the wind increased to moderate gale force (30 knots-plus) and stayed that way for the rest of our stay. It howled in the rigging 24/7, and the palm trees ashore thrashed around. There was no escape.

There were other yachts there. One in particular was a yacht from Perth.

John, the owner and builder, was a plumber by trade, and Perth to Cocos-Keeling was his first step on a circumnavigation. His yacht was also a Burrowitz design, a bit bigger than 'Dorado', and more of a motor-sailor. He was a guy who liked a drink and had a well-stocked locker full of duty-free goodies. He was a bit worried about taking his yacht over to West Island and leaving it there on anchor while he walked to the Passport Office. He

Chapter 4 - Circumnavigation

badgered me to come with him, and I did go in the end. Apparently, a couple of years before, there had been a yacht which had anchored there, and been left while the crew had gone ashore. While they were gone, the anchor had dragged, and the yacht had gone out through a pass in the reef and disappeared, never to be seen again. It had been a news story in Perth and was a cautionary tale for anyone.

Another yacht that was there was a French sloop with three young army officers on board. They were delivering the yacht from Tahiti to the south of France for a senior officer. Nice guys, all of them, and highly organised. They would fish nearly every day, and barbecue ashore. Everybody was invited, and they'd open a bottle of delicious French wine. We were sad to see them go.

There was another boat there, I hesitate to call it a yacht. It was a Chinese junk design, with a ferro-cement hull. There were a group of Americans on board who claimed to be doing ocean research sponsored by an American university. I suspect most of the research that they did came out of a bottle. They didn't socialise much with the rest of us, but there were two gals there, who couldn't have been more than 20, and would come around when there was a BBQ happening. They were unhappy with their situation, and wanted out of it, and back to America. They checked John out, and I knew he was looking for crew, but nothing came of it. They also checked me out, but I didn't fancy them at all, and I wasn't going their way. Eventually they all left, and I was on my own.

The trade was blowing as hard as ever, and I hung on for a few days hoping that it might ease off even for a short time, but I was getting lonely sitting there on my own. This was no good and I knew that I was going to have to bite the bullet. First, I had to go over to West Island to get my passport, which was going to take most of the daylight hours of one day. So, up early in the morning, a cup of coffee, and into it. Get the anchor up, and off across the bay. I'd done the same thing with John 10 days ago. There was plenty of coral around, but the water was quite deep.

It was only when we reached West Island that the amount of coral increased, and it was obvious that we could go no further. Bearing in mind what had happened to other people, I had decided to use both anchors this time. There was about 10' of water under the keel, and the holding didn't look too bad, but the wind was whipping across the bay. I launched the dinghy and checked the anchors from the surface. The water was clear, and I could see the anchors clearly. They both looked OK. I had also let out double the length of chain recommended for that depth of water. What could go wrong? I picked up the paperwork that I thought might be necessary, got into the dinghy and rowed ashore, pulled the dinghy up above the high-water mark, and off I went down a narrow road.

It was about four miles down this road, which took me about an hour and a half. I was starting to wonder when we would ever get there, when all of a sudden it all opened out and we were there. Buildings, the start of an airstrip, people standing around, and what's that? A huge aircraft was standing near the end of the runway. I started walking towards the buildings. Which would the Passport Office be in? There were two guys walking nearby, and I asked them. Right there, sir, they said, in that building. Just go around the corner. Young guys, fresh faced, crew-cut, uniformed, American accents. I went around the corner and found the Passport Office.

A fellow was at the desk, and I could see others in an office. I introduced myself and asked for my passport.

Chapter 4 - Circumnavigation

'Mr Brown,' he said, 'we've been expecting you. All the others have gone.'

'Is this part of Australia now?' I asked him.

'West Island operates as an Australian territory,' he said. 'The rest will follow soon.'

What were the Americans doing there, I asked him?

'Having a 12-hour rest period,' he said.

'D'you think the Russians know about it?' I said.

'It's not the Russians we have to worry about,' he said. 'It's the protesters. Can you imagine flying one of these things loaded with nuclear warheads into Melbourne or Sydney?'

Yes, I could just imagine, I said.

'Where's your next stop?' he asked.

I told him I would head to Mauritius. Good luck, he replied.

'Just one more thing,' I said. 'Is there a shop here that sells fresh bread or eggs, etc?'

'No, there isn't,' he said. 'But if you go along next door, you will find a café there making hamburgers for the Americans. They might do one for you if you speak nicely to them.'

So, I tried my luck, and I got two huge bacon and egg rolls, on the house. God bless America!!

The road back didn't seem so long, and as I approached my eyes were searching for 'Dorado', and there she was, riding easily, no problem. Launch the dinghy, get in and start rowing.

It's only about 100m out to 'Dorado', but we're going straight into the wind this time. I'd only just left the shore and was just settling into my stroke when there was a bump, bump, bump, coming from under the dinghy. What the hell is that? I didn't see it straight away and started rowing again. Bump, bump, bump. This time I caught a glimpse of it. It was a hammerhead shark, trying to tip me out of the dinghy. Shit!! The problem I had was that every time I stopped rowing we started going backwards towards the passage through the reef. The only option was to row and keep rowing. The shark, which was about 6' long, fitted nicely under the dinghy. It kept trying, but wasn't big enough to do any serious damage, it kept trying though, and it took me a good 10 minutes to row out to 'Dorado' against the strong wind. I was puffing and blowing by the time I got on board, and got the dinghy secured.

I then had to start on the anchors, which was another heavy job. I made that easier by starting the engine and putting it in slow ahead. By the time we'd gotten back to Direction Island, it was late afternoon, and it was an easy decision to say we'd done enough for one day. A good night's sleep, the last I'll get for a while.

In the morning, I was up early and got the yacht ready to leave. This was going to be a rough trip, and everything had to be stowed away just so. It was 2460nm to Mauritius, and I was hoping to do it in 20 days. The course was about west-sou'-west and that meant we would be reaching for most of the way, it would be wild and bumpy and wet. The sails were all ready to hoist as I hauled up the anchor and stowed it safely below. We motored through the pass, and as soon as we cleared the tree line, and we started to feel the force of the wind, I hoisted the sails. I had already put two reefs in the main, and set the #3 jib, and after I hoisted them, and trimmed them for a west-sou'-west course, the next thing was the Aries. I was going to relying greatly on the Aries on this voyage, and it didn't let me down. Not in my wildest dreams could I have contemplated such a voyage without it.

As soon as we were clear of land, it became clear that the

Chapter 4 - Circumnavigation

wind was sou'sou'-east rather than sou'-east, and the strength was 30 knots-plus, which meant there was a fair bit of sea running, with spray flying. 'Dorado' was doing seven knots even under reduced sail. Taking navigational sights was interesting to say the least. Wet weather gear with the harness connected to a strong point was essential. Finding a horizon was guess work, but I got it done.

Below decks, I filled in the log without fail, but living was very basic. Rest came in little grabs, sitting on the bunk fully clothed, ready to react to any change. Even in the cacophony of noise caused by the wind howling in the rigging, and the boat crashing through the seas, I would know instantly if something went wrong. The course needed to be checked constantly, sail settings needed to be checked and adjustments made, the Aries needed to be checked constantly, too, and adjusted if need be.

This went on for 14 days, when the conditions started to ease ever so slightly, the wind moved to sou'-east, and the strength dropped below 30 knots. On day 16, we passed the island of Rodrigues, which was a good navigational check. I'd thought of stopping there but decided not to in the end. The harbour is very small and restricted, and Mauritius is only three days away. The breeze then dropped away completely, and we were becalmed for a while before it came back from the nor'-east at 10 knots and helped us to cruise into Port Louis on the 20th day.

We went in as far as we could until we found a vacant wharf and tied up. I had hoisted the yellow practique flag, but no one seemed to be paying any attention. I'm going to have to go and find them myself, I thought. The place had an iffy feel to it. Don't be here any longer than you have to, and don't leave your boat unattended. Nobody was coming to check my papers, so I was going to have to go looking for them. As it happened, they were nearby, and quite pleased to see me.

Mauritius had been both French, then British, until independence in 1968. I never saw much evidence of British influence, the culture was mainly French, and the spoken language was French. Visiting yachts rarely stayed around Port Louis. They mostly headed up to the northeast corner of the island, to Grande Bay, and that is where I headed. It is a bit of a day sail up the coast. I found Grande Bay a very welcoming place, a safe and secure anchorage, and ashore was a hotel (Hotel de Paris), which backed onto the bay, where there was a bar, and showers which visiting yachties were welcome to use. Just down the road there was a boulangerie run by a mother and daughter, where yachties were very welcome. They were often at the hotel bar, and one night they got a group of us together, and we headed off to the Club Med. Club Med had a very high profile in these days, and entry was very restricted, but the girls knew somebody, and we managed to get in. A good time was had by all.

I'd been there for about two weeks, and I'd been expecting John to turn up, but he hadn't showed, and I felt it was about time I got going again. In fact, I missed him by a couple of days. He'd stopped at Rodrigues, before coming to Mauritius. I often reflect on missing these places. You get one chance, don't you? My plan was to go direct to Durban, a voyage of about 2000nm. The course was approximately south-west which took us just to the south of the French Island of Reunion. I wasn't planning to stop there, but maybe I should. I met people later on who'd had a great time there. Anyway, after passing Reunion, we have to pass south of Madagascar (yachts are not welcome in Madagascar and are advised to give it a wide berth as there have been reports of piracy).

I left Port Louis and headed south-west, but we were in the

Chapter 4 - Circumnavigation

wind shadow of Mauritius, and it was the next day before the trade wind returned, and we settled in to do some sailing. Again, the trade wind was more sou'-sou'-east than sou'-east but was not as strong 15-20 knots. It was going to be a close thing; could we pass south of Reunion without tacking? I checked the chart very closely. There didn't seem to be any dangers, but it was going to be after dark before we cleared it. I was hoping that the breeze might have shifted a little to the sou'-east during the day, but no such luck. I held the course until we were quite close, I could see cars running around a roadway. Tack, we might have made it, but the stress was too much for me, go on the starboard tack for an hour, no big deal, then back on the port tack, bye bye, Reunion, I should have gone there, even for a week. Too late now.

After we'd rounded Reunion, the breeze did shift to the sou'-east and held for a few more days, and we moved along quite serenely. I knew that the trades wouldn't last much longer, and once we got down the

African coast, we'd find a strong current, and different weather systems.

One day I was on deck and some movement caught my eye, I looked more closely, and lo and behold! There's a big fish under the boat. On closer inspection I could see that it was a swordfish, and if you included the sword, it must have been 15' long. I'd read stories about yachts being attacked by swordfish, and I didn't want it happening to me, not by this monster anyway. Just leave it alone was the best solution, it seemed to be quite happy just moving along with us in the shade, and maybe also the security.

It was on this voyage that an unusual navigational situation arose. I was doing the usual sun sights during the day, which is quite sufficient for accurate navigation, but there was an opportunity in the evening to check the day's results. It so happened that the planet Venus was an evening star, and after sunset there it was bearing due west. Also, the navigational star Vega was high in the sky to the north, so after I did the shots and plotted them on the chart, we had a little right angled cross marking our position on the chart.

I had decided that I was going to try and reach the coast at Richards Bay, about 100nm north of Durban, then use the strong current to help us down the coast. There was a strong navigational light at Richards Bay and that would be helpful. I knew we would be reaching the shipping lanes soon and I knew there would be little rest from then on. The last evening shot that I got had us around 100nm east of Richards Bay. The breeze was fading fast, and I thought I'd just get a few hours' sleep if I could.

I was up and about before dawn, and what a dawn it was. The sky had clouded over, and the sun came up a lurid red, and the clouds became a reddish-brown, the sea had turned lumpy, which told me that there was a strong wind somewhere, and we'd be getting it soon. I started up the auxiliary to give us some stability, and dropped the jib, but kept the main hoisted, which also helped with stability. About two hours later, a faint breeze came in from the nor'-east and started freshening, and before long we were bowling along under sail (full main and #2 genoa). The seas were lumpy still, and we were possibly entering the strong current that runs to the south-west down the African coast. The breeze kept freshening, and soon I was reefing. Late in the afternoon, I started to see shipping to the west of us, and it was obvious we were entering the shipping lanes. The breeze had now increased to over 30 knots and I had to put the second reef into the main and change the jib down to #3. Our course was due west, which meant we were cutting straight through the shipping

Chapter 4 - Circumnavigation

lanes. We were also being set south by the current, and that meant we would be unlikely to see the Richards Bay light.

About an hour after dark, and it was time for my evening meal. But not tonight. There was shipping everywhere, and I couldn't risk leaving the deck. I had my navigation lights turned on. I also had a device fitted near the top of the mast, designed to reflect radar signals and show up on the radar equipment of other shipping. It had a good reputation, but would it work tonight, with shipping everywhere, and in near gale conditions and big seas?

After a while, things seemed to settle down a bit. There were ships around, but they seemed to be well clear of us. Time to have something to eat, just something simple, maybe warm up a can of soup, then a cup of coffee. It wouldn't take long. I was busy over the stove, something made me look up. Oh No!! One of the ships I'd been watching had changed course and was now coming straight towards us. I had to tack immediately. It was close. The ship passed about 100m away, a hairsbreadth in these conditions. It was rolling from side to side as it went through. I could see two guys on the bridge looking at me, I gave them a wave, and they waved back.

It took me some time to settle down. I got back on course, and we seemed to be through the shipping lanes. I could still see the lights of ships in the distance, but they were behind us now. We were still some distance off the African coast, but I wasn't sure how far. I was a bit shaken by the incident and the weather and the confused seas. I stayed on deck all night just in case we encountered another ship or fishing boat, but nothing happened.

By dawn, the wind had eased slightly, and I shook a reef out of the main. I had read that on this coast the nor'-east blow was followed by a sou'-west blow, and I wanted to be in Durban before the sou'-west blow arrived. It was still cloudy, and it seemed unlikely that I'd be getting any sights this morning. That meant that I was going to have to rely on DR (dead reckoning), and the current would be a big factor. I adjusted the course to west-sou'-west and calculated that we had around 50nm to go to Durban.

The conditions continued to ease; the cloud was thinning, and around the time of the noon latitude reading I thought I might get a sight in. The sun was behind thin cloud, but I did get a good sight, and it showed that we were getting close to Durban's latitude. I grabbed the binoculars, and got back on deck, and sure enough through the haze I could see the high bluff of the Durban coastline. I had to harden up the sheets to lay the entrance to Durban harbour. The current had had a much bigger effect than I had estimated, and we would have missed Durban but for the noon lat.

There is an area inside Durban harbour where new arrivals have to wait for clearance, and I anchored there late in the afternoon showing the yellow 'Q' flag. It was obvious the customs were not coming today, and I was exhausted, so it was straight to bed. Sometime during the night, I was woken by the wind. It had swung to the sou'-west and was blowing very hard, indeed. I got up and checked the anchor. It seemed to be holding OK. I veered an extra few feet just in case, and went back to bed, thinking we were lucky we were here and not out to sea on a night like this. The officials came in the morning, and checked that my vaccinations were up to date, I had no drugs on board, no dirty magazines, no flora or fauna, my passport was up to date. They were very thorough, but fair, and I had no problem

Chapter 4 - Circumnavigation

with that. When they were about to leave, they told me to go right up to the top of the bay to The Point Yacht Club where I'd be given a berth.

They also told me that where I'd anchored was foul ground and I might have trouble getting my anchor up. They stood by whilst I tried to get the anchor up, and sure enough it was stuck, and I couldn't move it. They moved in with their launch, which had a powerful winch on the bow, and tried to lift it. More and more pressure was applied, until all of a sudden something gave, and the chain started to come in. I've lost another anchor for sure, I thought, but no, it was something on the seabed that gave way, and the anchor was OK. Thanks very much for your help, I called out, as they motored off.

In 1978, apartheid was still the rule in South Africa. Many countries had sanctioned South Africa, and the Afrikaaners resented that and were suspicious of people like visiting yachties. There were four main groups in the country, namely Afrikaaners, who were the biggest group, then the English speakers, who were mainly business people, including miners, etc, then the Indians, and the black tribesmen (Zulus and Zhosa). The main language is Afrikaans, followed by English, but it does depend on where you are. Durban is a place where English is spoken widely. At the Point Yacht Club, the main officials are all Afrikaaners, but there is a special appointee who deals with the visiting yachts which come from September each year, up till about Christmas. Bob Rogers does a wonderful job looking after all these diverse people. In 1978, there were 36 yachts, the majority of which were American: nearly 100 people in all, each with their own needs and desires.

Another situation in 1978 which affected us all was Rhodesia. It was on its last legs. Mugabe was gaining ground at an ever-increasing rate, and white people were just leaving everything and getting out. A lot were coming over the border and into South Africa bringing whatever they could with them. South Africa was letting them in but wasn't letting anybody go north.

When I got to the yacht club, it was Bob Rogers who I dealt with, and I liked the guy straight away. He showed me all around the place and made me feel welcome There was a nice bar which looked out over the marina, and a separate restaurant (not on my money). Downstairs, there were showers and a laundry, a wharf where you could tie up your dinghy. We were close to the centre of the city, and there were plenty of shops nearby. We were advised not to go near certain parts of the city. I told him that I wasn't here to cause trouble, and I wasn't a protester and so on. That's good, he says.

'Just understand that sometimes trouble comes looking for you.'

'I hear what you say, been there and done that,' I told him. And so, I settled in, getting to know people, and doing a bit of socialising.

One day, Bob took me aside.

'D'you want to earn a bit of money while you're here?' he said.

'Yeah, tell me about it,' I replied.

There were some people from Rhodesia who had built a yacht, put it on a truck, and brought it down to Durban and launched it, and fitted it out. Now they wanted to get it to the Seychelles and charter it. Their problem was that they were not experienced yachtsmen and they needed someone to skipper the yacht up to the Seychelles. Would I be interested? Yeah, well, I told him I'd be interested in talking to them anyway.

Chapter 4 - Circumnavigation

Next morning, I met three people, John, the son of the owner had been a schoolteacher; Karen, John's girlfriend, was Swedish, but her family had come out to Rhodesia to farm; and Jean Pierre, who was French, whose family also had come out to Rhodesia to farm. The owner of the yacht, John's father, was still up in Rhodesia. He'd been a spitfire pilot during World War II, and he, too, had come to Rhodesia after the war to farm. He was planning to join them in the Seychelles after they were set up. We had a chat for a while, and a beer or two, and then went to look at the boat. It was a 52' cutter rigged sloop of ferro-cement construction. I had a good look over it. All the sailing gear looked adequate and was new. The anchor and chain were all new, with a manual winch mounted on deck. Below decks, the fit out was fairly basic but adequate, and there was a big powerful auxiliary. There was a depth sounder and a VHF radio. Durban to Mahe, was a voyage of 2000nm, and with a crew of four I felt we could manage that easily. We're going for a sail on Sunday John said, would you like to come with us? Sure, I said, I look forward to it.

They have this rule in South Africa that you can't take your yacht out to sea unless you are an experienced, accredited, sailor. Of course, you can't get the experience unless you go out sailing, so it is a kind of catch 22 situation. John and his friends had no offshore experience, so they needed someone like me to be there with them. As an overseas visitor, I was exempt from the rule. Sunday was a perfect day for a sail, sunshine and a moderate nor'-easter. We tacked up the coast for a while, then turned around and reached back into the harbour. We passed another yacht heading towards the harbour showing a yellow 'Q' flag, and there was a big guy with a full beard on the tiller who waved to us as he went by.

The day was a success for us, and when they asked me if I'd

skipper them, I said I would, depending on the money. A discussion ensued at the end of which it was agreed that I would be paid 500 rand (= $A500), and an airline ticket back to Durban.

'When are you going to leave?' I asked. Sometime in November, they said.

'The cyclone season in the Mozambique channel starts in November,' I said. 'I think you should get going as soon as possible.'

We left it at that for the time being. They said they would load stores for 2-3 weeks. We agreed to meet again at the end of the week to check on progress. I urged them to get ready as soon as possible and offered to help them if I could.

Next day, I met the guy we'd seen arriving when we were out sailing. His name was Val Howells, a giant of the yachting world. He was a friend of Sir Francis Chichester, and took part in the first trans-Atlantic, single-handed race in a yacht that he built himself. A larger-than-life figure, I had come to know him quite well by the time I left South Africa.

We sailed out of Durban on November 1, later than I'd have liked. The good thing was we had a favourable wind sou'-west breeze of 35 knots, and we made great progress for the first three days, by which time we were well into the channel. As we went north, the breeze gradually eased and shifted to the sou'-east. They told me that they wanted to stop at Mayotte (one of the Comoros islands). There had been some talk about this before we left, and I told them if they wanted to do that, then they had better get the appropriate chart. They couldn't get the chart but had gotten hold of the sailing directions. I'd been hoping that we wouldn't stop there, I wanted to get on with the job, but

Chapter 4 - Circumnavigation

something happened, which made it essential that we go there.

The breeze had dropped away altogether, and we'd started up the auxiliary to keep us going forward. It was Karen's watch on the wheel, and she kept saying that the steering felt funny, and something was wrong. Turn the wheel from side to side and nothing happened. The steering was broken!!! Consternation!!! Then, almost as if he was expecting it, John said, I think I know what's causing that. We got down below to have a look. What a contraption, the words 'Heath Robinson' come to mind. There were even universal joints from a London bus involved. Right in the middle of all this stuff was a stainless-steel ¼ inch pin, which held the steering system together and, needless to say, it had sheered through. Now I discover that on its trip down from Rhodesia to Durban, the yacht had fallen off the truck, and there had been damage to the hull, and suspected damage to the steering. I was thinking, what the hell have I gotten myself into here. They showed me where the hull had been repaired, and it certainly looked OK, but what about the steering?

The broken pin really needed to be drilled right out and replaced, but that would be difficult in the open ocean, where there is always some movement, however slight. If we could find some way to jam the broken pin, long enough to let us get to Mayotte… So, we tightened up the broken pin, applying strong pressure in the hope that it would jam and got moving again. It held, but for how long? We were around 200nm from Mayotte, the sea was calm, and there was no wind to speak of. Just be gentle with the wheel, very gentle.

Mayotte is a continental island but has a coral reef round it. There are passes through the reef, and one has a marked channel for commercial shipping, and that was the one we were going to try for. The weather remained fine, and we found the pass without any trouble, and followed the marked channel for 10nm before anchoring at the commercial port.

The Comoros was a group of French islands, but there had been trouble in the recent past and Mayotte had stayed with the French whilst the others had broken away altogether. I then discovered the reason why the crew were so keen to stop at Mayotte. Jean Pierre had actually been born on Mayotte, where his father had been a farmer. When the troubles had begun a few years earlier, the family had had to leave, and had gone to Rhodesia, where the same kind of crops could be grown. He was keen to make contact with some old family friends, to see how they were going. Jean Pierre wasn't allowed much time for this, as when we went ashore to see the officials, they took one look at the Rhodesian passports and told us to leave immediately. John told them about our problem and pleaded with them for a few days at least. They allowed us two days but refused us entry. It was tough being a Rhodesian citizen at that time.

John got busy with the steering. He managed to drill the broken bolt out, and lo and behold he had a spare. Was that a co-incidence, or were they expecting trouble with the steering? I told him that we would probably get to the Seychelles OK, but he had a long-term problem there that would have to be addressed.

The next morning, Karen and Jean Pierre snuck ashore and got some lovely fresh bread, and eggs, and a few other things, and we were set to go. We didn't have a weather forecast, but the weather had remained fine, and we were expecting to pick up the sou'-east trade wind soon after we left. It was around 800nm to Mahe, and the course was nor'east. There were potential dangers along the way, but as long as we kept to the course, everything should be OK.

We left Mayotte that afternoon with enough time to get

Chapter 4 - Circumnavigation

through the pass, and clear of any dangers before dark. A light nor'-east breeze had sprung up which rather surprised me as we were very much in the sou'east trade wind zone. The nor'-east breeze forced us off our desired course, and I was having to pay close attention to navigation.

Fortunately, it didn't last long before it shifted to north, which allowed us to get back on course.

I noted that the barometer had dropped by 2mb, which again was unusual for this part of the world. By dawn, it was obvious that something was up; the barometer had dropped again, and the wind was freshening. A low had formed to the nor'-west of us and was probably going to move to the south down the Mozambique channel. The channel was about to have its first cyclone of the season. At the moment, the cyclone was still developing, and the wind in our area would probably increase.

During the day, the breeze increased to 30 knots, and shifted to nor'west. The sea was rising, and that was having an effect on my shipmates. It was time to put a reef in the main, and I wanted volunteers to help me do it. Jean Pierre took the wheel, but nobody else would come on deck. Probably just as well that I did it myself; nobody else had a safety harness. I'd brought mine with me just in case, and I certainly needed it that day. The jib was a roller-furler, and reducing sail was easy with that.

The sky was partly overcast, but I managed to get my navigational sights, which was important. I hadn't brought my sextant with me. John had an old one, which the owner had given to him. It was out of adjustment, but I soon fixed that. The mirrors were OK, and it was good to go. John also had a nautical almanac, which suggests to me that he was thinking of doing the navigation himself. The sights were good, and showed that we had made some good progress, and had passed most of the dangers.

The barometer was still falling, and the wind was slowly increasing. Night was approaching, and I felt I should put the second reef in before dark, and just as well. The wind shifted to the west at 40-plus knots, and we were on a broad reach, and running off the seas. The steering was OK and seemed to be standing up to the rough conditions. The barometer had steadied, and after 5-6 hours at 40-plus knots it began to ease. By dawn, we were rolling around in left-over seas with a light sou'-west-south breeze, and the barometer had risen back to where it should be. We reset the full main and started up the auxiliary and continued on our way.

It had been an exhausting 36 hours for me. I needed to rest. I left them with the wheel. Just keep steering the course, I told them. Wake me only if it's an emergency.

They got me up in four hours' time. They could see the palm trees on a coral atoll. Is that Mahe? Are we there yet? No, it's not Mahe. We're at least two days from Mahe.

The sea was settling down, the breeze had shifted to the sou'-east, mainly light at that time. With luck, it would pick up later on. The sun was out, and it was time for the morning shot. That didn't take long, and I stuck around for the noon lat. I plotted the results on the chart and showed them where we were and where we had to go. I showed them the island that they'd seen and pointed to another one that would appear shortly on the port side. We stopped the engine and trimmed the sails a bit. She was making four knots in the light breeze. It was a glorious day. They were doing one-hour turns on the wheel. Just keep watching the compass, and stay on course, was the message.

'I'm going back to bed,' I told them. 'Wake me up for the afternoon shot.'

Chapter 4 - Circumnavigation

It was pleasant sailing. The breeze had increased slightly and was steady at 10-12 knots, and we were cruising along at five knots. The situation continued for the next two days when the island of Mahe appeared over the horizon. We went straight to Victoria, the port, and main town of the group, and anchored close to the centre of town, where several other yachts were anchored. Clearance was the first priority, and that was achieved with no problems.

The Seychelles had become an independent nation in 1976 and all passports were welcome. French was the spoken language ashore, and Jean Pierre was in his element. The first day was just a celebration, they'd never done anything like this before, and just wanted to let their hair down.

Next morning, it was down to business. They had brought the yacht to the Seychelles because they intended to charter. Before they left Durban, they'd been given names and addresses of people to contact; people who could advise them, and help them to set up a charter business, and they were going to hit the road running. I went ashore with them, but my mission was different. I had to find a travel agent and arrange for my flight back to Durban. There was a café close to where we get ashore, and we agreed to meet there later on.

I found a travel agent without much trouble and showed him my ticket.

It was with British Airways (BA). They'd tried to get me on a South African Airways (SAA) flight direct to Durban, but they were booked out right over the Christmas period, and the only alternative was BA to Nairobi, change flights, then BA to Joburg, then SAA to Durban. A bit of a roundabout route, and even then, I was going to have wait for 10 days before they could fit me in.

I went back to the café to wait for the guys to come back.

There were some people there whom we'd met briefly the day before, and a conversation started up. There were three of them, and they were delivering a three masted sailing ship to Singapore where it was going for refit. They needed extra crew, especially a navigator. It's the kind of thing I would've loved to do, but 'Dorado' is sitting in Durban waiting for me to come back, so Singapore just wasn't going to happen. But they were good company, and it wasn't long before my shipmates returned.

We stayed in the café for lunch, and some other friends turned up. This time it was Di Di, and Fa Fa. I'd heard about Di Di, and Fa Fa in Durban, they were French brothers who owned a trimaran which they used to sail down to Madagascar to rescue French citizens who'd been stuck there. (Madagascar had been French, but in recent times the French had been overthrown, and the country had become independent in 1960, and more or less closed down to the rest of the world). Funnily enough, the phone system was still working, and people could still call up and make arrangements for someone to come and pick them up at some suitable beach. It is extremely dangerous, of course, and costs big $$$$$'s. Fa Fa reckoned they only accepted gold or $US. The brothers had become notorious and some kind of folk heroes. A similar situation existed in Cuba, where US citizens would run the gauntlet in fast runabouts from Florida to Cuba, pick up people wishing to escape, and return for rich rewards. We stayed in the café until late and had a meal there, and a good time was had by all.

The next day, my crew were going to meet some other contacts, and I went ashore with them. They were obviously busy, and I was going to have to find something to do to cover the next nine days till my flight was due. Get on the bus and go over to the beaches, someone suggested. All the resort hotels

Chapter 4 - Circumnavigation

were over that way, and they all speak English. It wasn't very far I could nearly have walked it, but it was like another world. Victoria was an industrial port, full of little bars and cafes. The language was French, and the culture was French. The other side of the island was all about resorts, and tourists, and English seemed to be the main language.

I walked onto the beach, and some activity caught my attention straight away. There was a power boat which seemed to be towing a parachute and hanging from the parachute was a human being. The power boat ran around for a few minutes and then gradually slowed allowing the parachute to slowly drop and deposit its passenger back on the beach. I discovered it was called parasailing, and it was very popular, judging by the queue of people waiting to have a go. About third in the queue was a little old lady with silver grey hair. Was she going to have a go? She sure was, and she dealt with it as well as any. Should I join the queue and have a go? I think I was too big and heavy, and they wouldn't have accepted me. That's what I told myself anyway as I wandered off down the beach.

There weren't many people around. I could see a lone figure sitting on a beach towel. What would she be? English, French, German maybe. Would she speak English. As I neared, she glanced up. Hello, I said. She looked at me with a funny look on her face.

'What are you doing sitting way out here all on your own?' I said. She gave me a funny look then said, 'Ahm mindin ma own business if that's all right with you,' she said. I was flabbergasted and could only stare at her. Finally, she said, 'So you're Scottish then'.

'Aye,' I said, sitting down. 'It's a small world isn't it.'

We sat and chatted for a while, then I walked her back to her hotel and had a drink or two, and a meal, and more drinks.

'I think I've missed the last bus,' I said.

'That's OK,' she said. 'You can stay with me.'

Next day, we got the bus over to Victoria and went into the café and met my shipmates. I had to go out to the yacht and get some fresh clothes, and at least make myself semi respectable. I might be away for a few days, I told them.

'You're not planning to leave here, are you?'

They told me they were negotiating a licence; they'd be around for a while. Jean, my new friend, was flying out two days before me, so we had a few pleasant days together. It was great for both of us, and we promised we'd meet up again in Scotland.

I just had time to get some washing done, and say farewell to my old shipmates, then it was on the bus out to the airport.

Back in Durban, there had been a number of new arrivals, and the yacht club was full of them. I'd loaned my dinghy to Val Howells while I was away, and I had to retrieve that first. He had a proposition for me. He wanted to get a few people together, and hire a minibus, and do a trip around Natal, game reserves, Rorkes Drift, the Drakensburg Mountains, etc. I was interested. I still had some rand left after my sailing trip, and that would be a good way to use it up. Two other people joined us, Charlie and Cricket. Americans, they'd been volunteers in the 'Peace Corps' in Southeast Asia for a number of years and had built a ferrocement hulled yacht for themselves, and here they were in Durban on their way back to the US. We tripped around for a couple of weeks, and it was a success. A couple of highlights are worth reporting. I think it was the third day that we drove to our second

Chapter 4 - Circumnavigation

game reserve. We arrived in the afternoon, and the game warden directed us to our accommodation, native style huts. But when you went inside you discovered they were fitted out with all mod cons. You could cook for yourself, or the warden would arrange for someone to cook for you. Before he left, he warned us, 'Don't go outside after dark. 'There are dangerous animals around here, and you could be attacked.'

We noticed that the floodlights had been switched on, and the whole camp was lit up. It was about 8.30pm, and we were on our third bottle of beer when a great roaring and growling started up. There were no windows in the cabin, but Val opened the door a little so we could get a glimpse of what was going on. Two adult lions had killed a Zebra and dragged it into the middle of the camp where the light was strongest, and they were ripping it to pieces. Other younger lions were standing around watching and waiting their turn. The show went on for some time, until they'd all had a feed, then the lions left, and everything settled down.

The warden came over to check on us and we invited him in for a drink. He told us about a place we should visit first thing in the morning. It was a water hole, where every animal and bird came to drink early in the morning. There was a hide, where a vehicle could drive into and be unseen, and we could watch the proceedings. Charlie and Cricket were keen to go too, so that was it. The warden gave us a map to follow, and we were up and about before dawn, and got down to the hide. Charlie got busy with his cine camera, and me with my Minolta. It was amazing to see all the wildlife that would normally be at each other's throats cooperating in this way for the benefit of all.

Cricket asked whether she could drive on the way back. Yeah, no problem. I asked her if she was familiar with the clutch and manual gear change. Yeah, she said, no problem, and off we went. All went well for a while. The track was well marked, and the country was fairly open.

Then we saw it. Standing on its own 250m-plus from us was a rhinoceros. Charlie wanted to stop and get some photos, and in a second, he's out of the vehicle and shooting. The animal turned towards us and lifted its head and started trotting towards us.

'Charlie, that's a black rhino,' I yelled. 'It's gonna charge us… Get back in the vehicle for Chrissake!!'

The animal was now at a full gallop and its intention was obvious. Charlie scrambled back in the bus, Cricket dropped the clutch—too quick—and stalled the engine. The air turned blue in the vehicle as I spouted every curse word that came into my head. The animal kept coming straight towards the back passenger door where I was sitting, closer and closer, I could see its beady little eyes fixed on me. I braced myself for the crash. No more than a foot from the door it turned its head and screeched to a halt then backed off for a few feet and stood glowering at us.

'Let's get out here, for Chrissake,' I said. Cricket turned round to me. Do you want to drive the vehicle, she said?

'No, no, please just go,' I said, 'and I apologise for all the bad language. I was a bit stressed out.'

Val was laughing, and Charlie was just staring at me. (I might point out here that I had rented the vehicle, and it was my name on the paperwork, including the insurance.) Back at the Warden's office, we told him the story. He thought it was funny.

'You were lucky,' he said. 'They don't always back off.'

Another highlight was the visit to Rorkes Drift, the site of a famous battle during the Zulu wars in 1879. Rorkes Drift was an outpost of the British army and had around 150 personnel, mainly Welsh. They were due to be relieved, but the relief was running late. A Zulu force said to be in excess of 3,000 warriors,

Chapter 4 - Circumnavigation

appeared and attacked the outpost. The defenders held on, and held on, until the third day, when the relief arrived, and the Zulus retreated. Only 11 of the original defenders survived, and all were awarded the Victoria Cross. Val Howells claimed to be related to one of the eleven, so this was a special time for him.

Charlie and Cricket were a strange couple. They both came from wealthy families, went to school together, university together, and were in the Peace Corps together, but they weren't married. Cricket was very accommodating to both Val and me. She seemed to think that the guys should be looked after. I don't know what Charlie thought of that, but he could be surly at times.

Back in Durban, Bob Rogers came to see me one day. There are some people up in the restaurant that would like to meet you, he said. It's Karen's mother, he said. I'd never been in the restaurant before, so I got myself tidied up a bit first. They invited me to sit down and have a coffee with them and were very nice to me. Apparently, Karen had written to her mother saying what a marvellous voyage they had had, and that they would never have made it if I hadn't been there. We spoke about the trip for a while, and how they hoped to start up a charter business. They were just very nice people, and I thanked them for inviting me up for coffee and excused myself.

It was Christmas by now, and my thoughts were turning to my next move—getting round to Cape Town. The locals recommended doing it in stages, and that made good sense. Durban to East London, about 255nm, and with the help of a strong current, should be less than two days. East London to Port Elisabeth, about 152nm, maybe just over a day. Port Elisabeth to Mossel Bay, about 154nm, and again just over a day. Mossel Bay to Cape Town, about 211nm, probably more than two days. The advice was to depart on the start of the nor'-east weather and arrive at the next port before the next sou'-west weather arrives. Rounding Cape Aghullas, and the Cape of Good Hope, the weather was more likely to be westerly, but this is the middle of summer, and it was unlikely that we'd see the strong blows that come in the winter months.

Right next to me in the moorings was another steel yacht. It had been built in South Africa by the owner, and he told me that he hoped to sail it to Australia one day. It was smaller than 'Dorado', yet the hull had been built in 3/16 plate. The mast was shorter than 'Dorado's, and just looking at it, sitting there in the water you could sense that it wouldn't sail well. The hot shot racers around Sydney used to look at 'Dorado' and say, 'Nice boat you've got there. It'll go a long way in a looong time.' I don't know what they would have said about this thing sitting next to me.

I hadn't seen much of him, but this was Christmas, and I started seeing more of them. He came down one day when I was there, and we had a conversation. We'd love to go out sailing he told me, but we can't, due to the stupid rules they have here. Could we ask you to come with us? That'll make it legal for us to pass through the heads. I probably should have checked up on him; gone to Bob Rogers and asked him what he thought. Instead, I was just the nice guy, 'Yeah, no problem,' I said.

'We want to go on Sunday,' he said.

'OK, see you on Sunday,' I told him.

Well, he turned up on Sunday with his wife and four kids, and enough food for a week. We got the auxiliary going, but it

Chapter 4 - Circumnavigation

was very small, and only pushed us along at two knots. We got outside and raised the sails. There was a nice, moderate breeze from the nor'-east, and we tacked into it for a while. Any sailing boat would do its best in such conditions, but this boat could not come within 50 degrees of the wind. He didn't seem to have any ability to steer a course by the compass. His daughters were far better than he was when I gave them a go.

When we went on the port tack and started moving away from the land, he started to panic.

'We've got to stay in sight of land,' he said. This was the guy who had told me he wanted to sail his yacht to Australia.

The day was disappointing, to say the least, and it got worse. I went up to the club to have a beer, and Bob Rogers pounced on me straight away. Have you got a minute he says and takes me aside.

'You've done the wrong thing going out sailing with that guy,' he says. 'He's barred from taking his yacht out. You should have come and spoken to me before you went out.

'Now certain people upstairs are upset,' he said, 'and I've been told to give you a ticking off.'

'That's OK, mate,' I said. 'I'm not offended. In fact, they are right, he shouldn't be allowed to go out especially not on his own.'

Bob asked me when I planned to leave.

'I just want this sou'-west weather to settle down,' I said, 'then I'll be off to East London.'

Bob seemed to hesitate for a second, and I said, 'In case I don't see you again, I just want to thank you for all your help while I've been here. I think you do a great job here, thank you very much.'

'It's easy to be nice to nice people,' he said.

The westerly weather continued to hang around, and I delayed my departure. I was in the club one day when a guy came up to me and introduced himself. We chatted for a few minutes, then he asked me whether I was available to crew on his yacht. There was a race next day, and he was shorthanded. This interested me. I'd never raced before on my yacht, or anybody else's, so I accepted. It was a tough race in tough conditions, and I was working the foredeck. The spray was flying almost continuously, and my eyes were stinging by the time we finished 12 hours later. We finished third, if that mattered, and were happy just to relax, and have a few drinks.

As luck would have it, the race saw the end of the westerly weather, and I was now ready to move south. The day after the race, the weather eased, and I raced around getting ready. Visitors to South Africa are required to check in with Customs at each stop they make, and that means more paperwork. I'd enjoyed Durban, but we were getting to the point where I was overstaying my welcome, and it was time to go. My eyes were still stinging from the salt spray during the race, but I wasn't worried about that at the time, just some fresh stores, enough for a couple of days.

Next morning, we moved out of our mooring berth and headed out of the harbour. A light sou'-east breeze greeted us as we headed sou'-west down the coast. Progress was a bit slow to start with, but the breeze freshened slowly, and became east 10-12 knots in the afternoon, and we were moving quite nicely. The coast is basically featureless, and there didn't seem to be much habitation. No lights were visible during the dark hours. We were inside the shipping lanes, and had no problems with them, although we could see their navigation lights not that far away. Fishing boats were not a problem either. It was just a pleasant sail

Chapter 4 - Circumnavigation

down the coast.

My eyes continued to trouble me. They were still stinging and were getting worse, if anything. Taking sights was getting difficult, and I was starting to think that I was going to have to seek medical attention.

After 24 hours, we had covered 120nm, which was quite satisfactory. I'd expected more, but we weren't getting much from the current. We probably needed to be farther offshore, but that meant getting out in the shipping lanes. No way! I'm very happy where we are, thank you. The breeze freshened a little more, and our speed increased to more than five knots. The seas remained fairly calm. There were no problems during the night. I had to stay awake, this close to the coast we could be wiped out by a fishing trawler.

About two hours before dawn, I started to see the 'glow' of East London ahead, and by mid-morning we were in the harbour tying up to the main wharf. There were a couple of American boats there that I'd met in Durban, and they greeted me, and helped me tie up. They noticed my eyes straight away.

'You've got 'Pink Eye', they said. 'You need to go and see a doctor.'

I was directed to the local hospital and saw a doctor, who turned out to be a Scot.

'Yeah,' he said. '" Pink Eye' is what it is. You need to rest your eyes for at least 10 days.'

He gave me a script for some drops to help them, but they needed rest. So that was me. Stay out of the sun; rest the eyes; use up the drops. After a few days, they were feeling much better. The drops were finished, and I was starting to think about the next leg, but the westerlies had come back, and the decision was easy. Stay where you are and rest up. The easterlies will return and that will be the time to go.

In 1979, East London was a city of 200,000-plus. There didn't seem to be any major industries. The harbour was based on the Buffalo River, which had been dredged out. Only small or at best moderate size ships could berth there, and it didn't seem like there was any major trade. The city itself seemed fairly quiet. There wasn't a yacht club that would welcome visitors, and the few pubs that I could find were hardly worth the name.

I'd been there over a week, and I was getting bored with the place. My eyes had improved and felt good. The westerlies had died down and were expected to be replaced with sou'-easters. It was around 154nm to Port Elizabeth. It was time to go. Paperwork done, a few fresh things and off. Outside the harbour there was light southerly, and a left-over sea from the westerlies. Not very comfortable for 'Dorado' and me. We persevered; we had no choice, and eventually the breeze swung to the sou'-east at 10 knots. Progress was a bit better then and I was expecting that the breeze might go more easterly and increase a bit.

Far from that. During the night, a change came through, the wind shifted back to the west and started to freshen. We were close-hauled, port tack taking us towards the shore, and starboard tack away from the shore. The breeze got up to about 20 knots, which made life interesting.

I was hanging onto the full main with #2 genoa. The coast was featureless, no habitation, no lights, nothing to warn the sailor. The chart showed no off-lying dangers, rocks, etc. A bit of cloud had arrived with the change and the moon had disappeared. I knew we must be getting close to the coast, and I was getting ready to tack anyway. A swell came from behind us, a little unusual, through the murk ahead I could see – a beach, and all of a sudden, I could hear the roar of the surf. Action. I let

Chapter 4 - Circumnavigation

the sheet go on the genoa and hauled the tiller over, we were wallowing between the swells. The main swung over and we were on the other tack, I was hauling in the genoa sheet like a maniac. I got it on the winch and left it there for a second while I reconnected the Aries, then back to the genoa sheet and trimmed it using the winch. Finally, I reset the Aries for the new course. We were moving away from the beach. The emergency was over.

I stayed on deck for a while watching and listening. Everything was OK. I should have tacked sooner. It's a bad habit, hanging on like that. One of these days…

I held the starboard tack until nearly daylight. The breeze shifted to the sou'-west and started to ease. I tacked to port and set 'Dorado' on course for Port Elizabeth. The breeze eased even further and shifted to the south. We were creeping along at two knots. Progress had been slow, 75nm in the last 24 hours. We were not going to be in Port Elizabeth tonight, that was for sure. I thought about starting the engine and decided no, we still wouldn't be in PE tonight. Just relax; we'll be there tomorrow.

I'd just done the morning sight and was below decks doing all the calculations when I thought I heard something. What was that? There it goes again. It was like a murmur. I have to investigate. I get up on deck and look all around—nothing. I was about to turn around and go back below when I saw it. A whale, snuggled up, right alongside us, on the port side. The murmur didn't sound like an angry sound, more like a contented, peaceful sound. What do think, I'm your mother? I shouted at it. It didn't answer, just kept murmuring. I couldn't see any other whales around, so it could well be lost.

It was as big as the boat. Were we in any danger from it? Probably not, unless the mother appeared and suspected someone was making off with her offspring. I'd been about to start up the engine. What would happen if I did that? I meant the animal no harm. I just wanted it to move along and get on with its life. I started the engine, and it took off immediately. Sorry about that. I let the engine run for an hour or so to charge up the batteries, then resumed our slow progress.

It was nearly dawn when the breeze shifted to the east and increased to 12-15 knots. That got us moving nicely, and we arrived in Port Elizabeth harbour in the early afternoon. Over in the south-west corner of the harbour, I could see a number of small yachts and I headed in that direction, anchored, and went ashore. There was a yacht club there, and I spoke to the secretary in the office. The club was open for sailing activities only on the weekend.

'It's OK to anchor where you are in the short term,' he said. 'No charge.'

I asked where Customs were, and the nearest shops. It was already obvious that Port Elizabeth was going to be similar to East London, and I wouldn't be staying very long.

The weather remained fine, and the breeze mainly light. The forecasts contained no warnings. It was late summer and strong weather was unlikely from any direction. PE to Mossel Bay was approximately 154nm. We did it in less than two days in variable conditions. We always had a breeze of some sort. Nothing too extreme.

Mossel Bay is a very attractive looking place, built on a hillside. The harbour is very safe with its protecting breakwater, and best of all there were a few yachts that I recognised. I was going to have some company this time. I hadn't seen Val Howells for some time. He'd become involved in a project in Durban and had invited me to be part of it. I'd said no, for various reasons, and Val didn't like that too much.

Chapter 4 - Circumnavigation

There was a handicapped couple in Durban, both confined to wheelchairs, and they had a yacht designed and built to suit their own unique circumstances and were determined they were going to sail this thing around the world. The mind boggles at the problems they were going to have to overcome, the first of which was to be allowed to leave Durban harbour. After Val became involved, a plan was worked out where Val would go with them and skipper the yacht round to Cape Town, after which they would be on their own. They must have passed me somewhere on the coast because, lo and behold, here they were in Mossel Bay.

There was also an Australian yacht, which I hadn't met in Durban, and two American yachts that I'd met in East London. It was nice to have some company for a change, to chat about this and that, and discuss tactics for the rounding of the Cape of Good Hope. Mossel Bay to Cape Town is a voyage of 211nm, and that would take us 2-3 days. Between Mossel Bay and Cape Aghulas, there are four headlands that stick out to the south and provide protection from westerly winds. The Americans, in particular, seemed very interested in these. My attitude was, I can pick the time of leaving port when the weather looks most suitable, but after I leave port, I'll take whatever comes. It is summer, after all, and if we do get a westerly it surely wouldn't be as bad the winter gales.

Cape Aghulas has a dreadful reputation, and deservedly so, but it is only 34.5 degrees south. After a few days in Mossel Bay, we received a forecast of moderate easterlies. That was good enough for me, and I made us ready and left the next morning. Away from the harbour and out to sea, we were greeted by a light to moderate sou'-east breeze. So, it was full main, and #1 genoa and we were off and running. The breeze lasted for most of the day, and we made good progress. Late in the day, the breeze shifted to the east and began to ease, and overnight it went to nor'-east then nor'-west and became light. I'd been motoring along for a couple of hours trying to make some progress, but an hour after dawn the breeze shifted to the west and started to freshen. Shut down the auxiliary, and set us off on the port tack, full main and #1 genoa. But not for long; replace #1 with #2 genoa, and soon after that we put the first reef in the main. The breeze was now gusting at over 20 knots and climbing.

According to my navigation, this tack would bring us up to Cape Aghulas. I confirmed this with morning sights and again with the noon lat. The breeze continued to increase, and I replaced the #2 with the #3 genoa. The sea was really starting to get up and every wave had a top on it and spray was flying everywhere. I held on with one reef in the main longer than I should have. 'Dorado' sails better and 'points' higher, and if you put the second reef in, you lose a lot of that. We were now starting to approach the coast, and there was the Cape Aghulas light house. No heroics today. We've confirmed our position; that's enough. Tack to starboard.

We made a long tack to starboard, too long, in fact, because we were now starting to get out into the shipping lanes. The wind gusts had peaked at 40 knots, and very slowly started to ease. More important than that, the breeze shifted to sou'-west, and that allowed us to reach all the way over to the Cape of Good Hope.

We were now about 65nm from Cape Town. The breeze was easing, and we were just trickling along. I would have been quite happy with that, but the number of ships was increasing. Great big oil tankers passing us, one after another. It was a nervous time, and I couldn't leave the deck for any length of time. One of the monsters passed us, then veered in front of us by 4-5nm.

Chapter 4 - Circumnavigation

Floodlights came on, lighting up the deck, and soon we heard the chattering of a helicopter. It landed on the ship, stayed for a few minutes and took off again. Change of crew? A crew member got injured? Chinese takeaways? I wasn't close enough to see. I decided to start up the engine to give us a more rapid reaction if an emergency arose.

There was no electric start system with my auxiliary, just a simple, old hand start. No problem if the battery goes flat. I turned on the water, grabbed the starter handle, and wound it up and dropped the valve lifter… nothing. Try again… nothing. Try again, a really good wind… nothing. It has never given me a moment's concern and started easily in Mossel Bay. I'm still using the same fuel that I bought in Sydney. Alf Taylor strongly advised me before I left Sydney to get some Easi-Start.

Guess what! What could have happened between Mossel Bay and here?

We've moved from the Indian Ocean and its warm current to the Atlantic Ocean and its cold current. Could that have made such a difference? I tried to start it again… nothing.

Well, I didn't have any Easi-Start, but I did have some petrol which I used in my small Honda generator. Diesel engines don't run on petrol, but if I got a drop of petrol on an old rag, and wiped it around the breather hole, would that be enough to get us going? It's worth a try. It worked; I cranked the starter handle as hard as I could then dropped the valve lifter. It went 'bang' followed by tic, tic, tic… We were off and running.

The breeze had dropped right away, so I dropped the jib but left the main up for stability. We were still 25-30nm from Cape Town, but I didn't want to hang around here in super tanker alley. It took a few hours and the weather remained calm for us to motor into Cape Town harbour.

I soon found the yacht basin, and there was someone there to meet us and allocate us a berth.

'Good morning,' he says. 'You kept going. All your friends are still east of Cape Aghulas sheltering. They were worried about you and couldn't contact you.'

'I don't have a radio,' I said.

'Aha,' he said, 'but you were spotted by the people at Aghulas lighthouse, and they reported your passing to us.'

'Thanks very much for your concern,' I said. 'It was those jolly super tankers that caused me the biggest problem.'

The yacht basin occupies a small part of Cape Town harbour and is quite crowded. Visitors are welcomed at this time of year but were allowed to stay for two weeks only. There is a slipping facility, which I took advantage of. It is a large slipway intended for much bigger vessels, but they use the four upright stanchions and allocate one yacht per stanchion. Each yacht is given two days to do what needs to be done, so you've gotta get stuck in. I needed to scrub back everything below the waterline, some minor paintwork repairs, replace the anodes, and two coats of antifouling.

It was hard work and dirty, but it got done, and back to the club for a good hot shower. The people at the club were a friendly lot, and they took me out and showed me around. One thing I wanted to do was get to the top of Table Mountain. There is a cable car, and I fancied a trip in that. It doesn't run every day, only when the wind is below a certain strength, and no tablecloth on the top of the mountain.

I was lucky. The day I did it, the weather was fine, even so there were gusts of wind about halfway up, and also on the way back, and the car was swinging around like billy-o. Table Mountain is approximately 3500ft high and offers spectacular

Chapter 4 - Circumnavigation

views in all directions.

Our two weeks in Cape Town went very quickly, and all of a sudden, I was racing around getting stores on board and planning the next move. This was quite simple. The island of Saint Helena lies about 1250nm nor'-west of Cape Town. There might be some sou'-west weather to start with, which would become sou'-east as we reach the trades. And that was the way that it turned out.

The only incident worth recording happened a few days out from Cape Town. I'd been below and came up on deck to check how things were going. I saw it straight away, a big black dorsal fin cruising up and down about 100m behind us. I realised that it was probably a killer whale, and the hair stood up on the back of my neck. It seemed to be checking us out, and I'm happy to report that after cruising around for a while it disappeared. The paddle of the Aries waggles around quite a bit, and that may have attracted it.

St Helena sits out on its own in the South Atlantic. It's not very big, but it's quite high. The anchorage on the northern side of the island is open and quite rolly and uncomfortable. Landing is difficult, the only place being where a short stub of a jetty with steps has been built. It is also open to the swell, and it requires nice timing to get your dinghy in the right position and stepping ashore without a disaster occurring.

There were several cruising yachts there, all of them I had met before, and we formed a bit of a community. There was a bus organised for a trip around the island to see the sights, which included 'Longwood', where Napoleon spent his last days. The population of the island is over 3,000, which surprised me a bit.

Apparently, it is a bit of a tourist destination. The pub, British style of course, was very popular, maybe just a little expensive. I stayed there for five or six days, and I might have stayed longer but the anchorage was very uncomfortable.

St Helena to Ascension Island is a distance of about 700nm, easy sailing in a moderate trade wind. The anchorage is open and even more rolly than St Helena. Landing is similar to St Helena, on a short concrete jetty, but even more care is required as the swells are larger and break on a beach just in front of you. I came a cropper the first time that I tried it, which was embarrassing.

There was a sort of general store near the end of the jetty, and the proprietor was very happy to see me. He was on for a chat and filled me in on all the dos and don'ts, mostly don'ts. Ascension has military facilities and other scientific facilities that are off limits to visitors. The island has a desert climate, and very little greenery. I think the general store was the most important thing for me.

The fellow in the general store ran a newsletter and he got my mother's address in Scotland and posted her a copy with a little story in it about George Brown and 'Dorado'. He also had fresh bread and eggs, which is fantastic for the sailor who has no refrigeration. I had some sail repairs to do, and some detailed planning for the next leg of my voyage. It would be our longest voyage to date, approx. 3500nm. We would cross the equator, then the doldrums, and the nor'-east trades, then a large area of variable weather. The Hiscock's had taken something like 55 days to travel that distance. Surely, we could do better than that. It was a challenge anyway.

I was on my own at Ascension Island. All my friends had headed straight for the West Indies, and after I'd done the sail repairs there was nothing much to hold me there. This was now

Chapter 4 - Circumnavigation

about the end of March, and we were on track to arrive in the South of England in June.

Ascension Island lies about eight degrees south of the equator. Six days sailing in a moderate trade wind, and we were almost to the equator. The breeze then faded out and left us becalmed. We were in the doldrums and very close to the equator. Let's not mess around was my attitude, let's get through the doldrums. Ever since Good Hope, the engine had been hard to start—just as well I'd gotten some 'Easi-Start' in Cape Town.

There was a big thunderstorm that coincided with crossing the equator. The rain came lashing down and everything got a good wash, including yours truly. Next day, a light breeze sprang up from the west, and we were able to take advantage of it until it died out in the late afternoon ahead of an evening storm. After the storm, everything remained calm, and I went to bed to get some rest.

Around 3am, I woke to find a light nor'-east breeze. Wow! Is it possible? Less than three days and we're through the doldrums. I rapidly set sail, full main and #1 genoa, adjusted the Aries… five knots on a course of almost due north… very happy with that. But there was something else, I could taste it and smell it.

Dust: the Sahara Desert was on the move, and it was going to South America. There was a lurid dawn, but the sun was largely hidden by the dust and when it came time for the morning shot, it was hopeless. I couldn't get a horizon. This problem continued all the way through the trade wind zone and only started to ease when we got to 23 degrees north (1380nm). I'd never read or heard of anyone who'd experienced anything like it. Visibility was no more than a few hundred metres, a potentially dangerous situation. To my knowledge there were no shipping lanes nearby, but you never know do you.

We'd come through the trades with the breeze about 20-25 knots. One reef in the main and #2 genoa took us all the way through. The trades continued way past the limit that I was expecting, and as we headed north, the breeze veered more towards the east and continued to blow at 20-25 knots. We never experienced any 'variables', or calms. As we continued north, the breeze slowly veered to the sou'-east, allowing us to lay the course for Horta, in the Azores.

A large zone of high pressure was responsible for our good luck, and we rode it all the way. We did Ascension to Horta in 31 days, compared to the Hiscock's who took 55 days. I was now 2-3 weeks ahead of schedule, and that meant I could rest up in Horta for a little while before taking on the North Atlantic.

Horta harbour features a long seawall, and visiting yachts tie up alongside the wall. It is not quite calm, and a small scend works its way along the wall. Good fenders are essential. 'Dorado's' fenders weren't adequate, and I had to go looking for something extra and ended up with some old car tyres.

Horta is a favourite stopover for people coming back from the West

Indies, and so there was plenty of company; yachts heading back to Europe after spending winter in the Caribbean. I became friendly with three young guys who were delivering a yacht back to the south of England. We hired a car and toured around the island. The whole area is volcanic, and there had been a small eruption just a few years previously.

Peters' café was a well-known meeting place for yachties, and I'd heard about it, and read about it. The first time I went there, I was sitting down having a beer when the garcon beckoned me over.

Chapter 4 - Circumnavigation

'Is your name George Brown,' he asked me, 'from the yacht 'Dorado'?' There was mail for me, he said, pointing to a board with letters on it. Sure enough, there was a letter there for me, but who was it from? I opened it up and all became clear. I'd been having a conversation with a couple of guys in the yacht club in Cape Town, and they'd been asking me about the kerosene stove on 'Dorado', and I told them how I could probably do with some spare parts, but I didn't have the time to go hunting them down. They told me they'd get me some, but I never saw them again. They must have gotten the spares and found that I'd gone and posted them to me c/- Peter's Café. I thought that was a very decent thing to do, and the very least I could do was to send them a card thanking them.

Hunting and killing of whales by the traditional method was still allowed in the Azores (1979) and in Peter's Café they showed us videos of this happening. It started with the spotters high on the cliffs, the launching of the chase boats, the crews rowing like mad trying to be first on the scene, then the kill, towing the carcass back into the harbour, and cutting it up. I'm against killing whales, but it was interesting watching it being done by people who have depended on it for thousands of years. It's part of their culture. You walk around the waterfront, and you recognise the faces of the crew. None of them are young, but they are all wiry fit looking guys, some with grey hair could even be in their 70s.

One day a yacht came motoring around the end of the breakwater. It was fairly big, probably over 50', aluminium construction, and was flying the French flag. It had no mast. They came right up the wall and tied up just in front of me. I gave them a hand to tie up, as you do. They'd come from the West Indies, where they'd been for the winter season. They'd been involved in some of the races that take place and had lost their mast when caught in a squall. The father had left the boat (reason unspecified), and the mother and three teenage kids were left to try and get the yacht back to France.

I was starting to plan the next leg of our voyage. I'd decided that I was going to go to Falmouth initially, and maybe up to Scotland later, depending on how we went. I'd decided on Falmouth because of the exploits of Robin Knox Johnston, who had started and finished his epic round the world non-stop voyage from there. It was about 1600nm to the nor'-east from Horta, and I was likely to get some variable weather, and possibly some strong weather on the way there. I was at the chart table one day pondering these things when someone came aboard. I stuck my head out of the hatch, and there was the young guy from the French boat.

'Bonjour,' he says. 'My mother would like to invite you over for lunch.

Would you like to join us?'

Wow, that took me aback for a second.

'Thank you very much,' I said. 'I'd love to come over. Just give me a second to freshen up.'

Lunch consisted of a few sandwiches and cakes washed down with beer. Conversation wasn't easy, her English wasn't very good, and my French was strictly schoolboy. The youngest daughter got up and left us, then the older one made excuses and left us. I was starting to think that I was overstaying my welcome, and when the boy got up and left, I thought I'd better get going too. She reached out and grasped my arm.

'M'sieur,' she said, looking me straight in the eye, so soon.

Chapter 4 - Circumnavigation

'Madame,' I said, catching up at 100mph, 'I'm all yours.'

She laughed, and that broke the ice. We got to know each other better, and 'lunch' was a regular occurrence. I stayed on a bit longer in Horta, and I couldn't think of a nicer place to stay.

The trip up to Falmouth took me 18 days. I left on a day when the breeze was light, and made slow progress, then a front approached from the sou'-west, and I had seven days of good breeze and made good progress. The breeze peaked at 40 knots as the front went through but otherwise didn't exceed 25 knots. The last few days, we had light variable winds, mainly sou'-east.

Approaching the English Channel was a worry. The traffic was very heavy, and I wasn't accustomed to it. I had to start the auxiliary several times to get us out of trouble, and I heaved a big sigh of relief as we entered Falmouth harbour and dropped anchor behind the moored boats. I was showing the yellow 'Q' flag, but nobody was coming near me. It looked like I was going to have to go ashore and find them. There was a guy passing me in a big launch, I waved him over and asked him. They won't come out to you he said, you'll have to go ashore and see them, he said.

'See that big building up on the hill,' he said, pointing. 'That's where they are.'

Thanks very much, I said.

'By the way,' he said, 'are you staying here very long?'

I told him I'd be staying a few days at least, maybe longer.

'I have to ring my family up in Scotland first, and then decide what to do.'

He pointed to a vacant mooring.

'See that mooring over there?' he said. 'That's my mooring, and it's not in use. You can have that if you like for the time being, no charge. If the harbour masters lot come around asking for money, don't give them any, OK.'

Thanks very much, I said. 'How can I thank you?'

'Don't worry about it,' he said, and he introduced himself.

'I run the fisherman's co-op just up the river there. When you know what you are doing, come and see me.'

I launched the dinghy and rowed ashore and left the dinghy where I could see other dinghies tied up Would it be safe? Hope so.

The hill was quite steep, and when I got up to the Customs House I was puffing and blowing. They were on the second floor, and quite pleased to see me.

'Thanks very much for coming,' one of them said. 'Would you like a cup of tea? We don't get many foreign yachts nowadays.'

In the Common Market, they said, 'free movement' meant they hadn't processed yachts for ages.

Processing was a formality, and a stamp in the passport was all I had to show for it.

Back in the town, there was a pub close to the dinghy and I stopped there and had a counter lunch and a couple of pints. Very nice, but a bit pricey, I thought. I tried to ring home, but no one answered. That was unusual. My parents were getting a bit elderly, and neither was in great health. I had another pint and tried again… no answer. I had a number for my brother, but it was back on the boat. Tried the number one more time… no answer. I went back on board to get my brother's number. Give it a while I thought, wait until the evening; there'll be a better chance of getting my mother.

It was after 6pm when I went ashore again. I tried my

Chapter 4 - Circumnavigation

mother's number first and got through straight away. The news was that my father had been taken to hospital that day and was very ill. When was I coming up? Of course, I was coming up. It would just take a couple of days to organise myself. Was I flying? I doubt it, but I'm going to ring Andrew, my brother. I told my mother I would call back the following night.

It was decided that train would be best, and Andrew would meet me at the station. First thing next morning, I went to see my friend Sam at the fisherman's co-op to explain what was happening. No problem, he said. Just keep in touch and let us know when you're coming back. And he added, I know it's a bit of a long row, but your dinghy will be much safer here than at the public steps, so bring it down here and leave it here whilst you're away. How lucky was I meeting that guy! I couldn't have done it without him. I was up in Scotland in the evening and went to the hospital the next day.

My father didn't stay in hospital very long. They said there was nothing more that they could do for him and sent him home. He lasted for another 10 days. There was the funeral of course, then I stayed on for a while. I couldn't just walk away and leave my mother on her own.

It was a rather sad summer altogether. Suddenly it was the end of August, and it was time to go. There was work that needed to be done before 'Dorado' could leave on the second half of our circumnavigation. I'd been in contact with my friend, Sam, at the fisherman's co-op. He would be expecting me. The main work that needed to be done was 1) get the boat on a slipway somewhere, clean the hull, patch up any bad spots, inspect the anodes, replace, if necessary, one or two, and if I have the time, coats of antifouling. Slipways are expensive and I was already conscious of a very tight budget, but this job had to be done. 2) I'd been aware for some time that there was a problem at the top of the mast. The main halyard didn't seem to be running truly and was tending to stick. I'd already been up the mast to inspect it. The main sheaf was worn out and was needed replacement. I didn't want to take the mast down for what was really a minor job but having to climb the mast and work at the masthead for half an hour or so was a tricky proposition. It had to be done whilst we were in harbour, where the water is calm. It would be a fairly desperate job in a seaway.

I had fitted steps to the mast, which was a wise move. I put tools and spare parts in a bucket and hauled it to the top of the mast, then up I go… harness myself to the top of the mast and get on with it. It doesn't take too long, and it's done. Check the halyard carefully… it seems to be OK. The main halyard comes under severe and prolonged stress, and it's out of sight for most of the time. As far as the slipping was concerned, Sam advised me not to try and book a slip. Come down to the co-op, he said. There was a hard stand there that I could use for two or three days, no charge. That's the best price isn't it.

Falmouth has a tidal range of 10'-plus, which is ideal for the hard stand. I got the bottom clean and two coats of anti-fouling.

3) The auxiliary is direct salt water cooled with a water pump to control the flow. There was a plastic valve regulating it. Occasionally, a foreign body would get into the system and block the flow. After the first time this happens, you know immediately what the problem is and fix it easily. Turn the inlet valve off; dismantle the water pump; and find the foreign body and remove it. Also remove and replace the small plastic valve in the middle of the pump; reassemble; start up the auxiliary and you're on

Chapter 4 - Circumnavigation

your way.

The small plastic valves were the problem. I needed a few spares; they needed to be replaced regularly. I tried to get some in South Africa, but no joy. There was supposed to be a 'Yanmar' agent in Falmouth, but I didn't know the address. Sam knew the guy. He wasn't actually in Falmouth, just a short distance down the coast, and we gave him a ring to make sure he had the parts. Yes, he did.

'Take my van,' says Sam. 'I don't need it this morning.'

This guy was just so good to me; I owed him a lot, and I'm sure he didn't realise how much.

There were still some canned food supplies left over from Sydney, but not enough to get me back, and I had to restock. Falmouth was the logical place to do it. Most places I would stop, I would be able to get some fresh supplies. I wasn't planning to stop anywhere for any great length of time. I was hoping to be back in Sydney about a year from now, so I didn't need all that much, really. A few trips to the supermarket were enough, and don't forget the duty-free liquor. Check all the charts carefully, check all the sails carefully, a few minor repairs. Compass was OK. It was set for the southern hemisphere, and we were now in the northern hemisphere, but it wouldn't be long before we'd be back in the south. I knew how to check it myself, and I did so nearly every day when we were sailing.

I made sure I had a copy of the Nautical Almanac for 1980. I'd be in the West Indies for the New Year, and I might not be able to get it there. The route we would take would be very similar to the Hiscock's whose books I had read and reread many times. From Falmouth we'd head for Baiona in Northwest Spain, then Madeira, then Tenerife in the Canaries.

That would be the jumping off point for the Atlantic

crossing, and Barbados would be the arrival point in the West Indies. Christmas and New Year in the Windies, then Panama and the Pacific.

By late September, I was ready to go. I thanked Sam for all his help and promised to send him a card from Tahiti. I went to see Customs again. They put a stamp in my passport and thanked me for coming. Under the common market, things were much more relaxed, they said.

'Don't worry too much about Spain and Portugal,' they said.

'In the West Indies, they get overwhelmed with yachts in the winter months, and don't want to know you.'

I took in all this information, which turned out to be largely accurate.

We left Falmouth right at the end of September, with a weather forecast of light-moderate sou'-east winds. The breeze finally arrived in the late afternoon as we were negotiating the shipping channels. The breeze held light all night and when we came upon the French fishing fleet, I had to start up the auxiliary. It was like dodgem cars… change course to avoid one, and another comes straight at you. It went on for hours, and it was amazing that we didn't collide with someone.

Now we were entering the Bay of Biscay, that notorious stretch of water.

But the weather stayed light, and we struggled for the next four days. On the fifth day, we got a nice easterly and started to make some decent progress. The breeze then backed to the nor'-east and increased to 30 knots, and gusting.

My navigation had been quite good, and we were approaching Cape Finisterre, which would be an important mark for us. Then, I would know that we were about 50nm from Baiona. The breeze then dropped away, and we were almost

Chapter 4 - Circumnavigation

becalmed off Cape Finisterre. I now knew that if we kept going straight south for 50nm we would be off Baiona in the morning.

Things became complicated overnight, when a thick fog set in, and visibility was down to less than 100m. With our own diesel rattling away, we couldn't hear much either. It was a nervous time, and luckily nothing bad happened. By daylight we had nearly covered the 50nm, and just when we needed it most the fog started to lift, and we could see land. It was Isla de San Martino, and we only had to adjust our course slightly, and head straight for Baiona. Around the headland with the castle on it, and into the bay itself.

The anchorage was obvious, there were a number of yachts there already, and I took my place among them. The most memorable of the people I met were two gals on a timber yacht. They were reporters and had been trapped in Iran when the Ayatollah took over. They had plenty of stories to tell and were good company. They were looking for someone to deliver their yacht into the Mediterranean and asked me if I could help them. It was out of the question. No way was I going to leave 'Dorado' swinging on an anchor in Baiona while I went off on some adventure.

There was a disco-nightclub ashore, and the yachties would meet there most nights about 5-6pm and stay till about 8pm usually. Spanish people wouldn't turn up until later, so we had the place to ourselves. The night before I left, there was a barbecue ashore. Someone got some shrimp, and I indulged in shrimp and red wine. I knew it was the wrong thing to do, but what the hell. I felt decidedly off-colour when I got up the next morning, but I left anyway, thinking I would improve during the day.

The weather couldn't be more favourable. We experienced

the 'Portuguese Trades' all the way to Madeira. I knew little about it, especially the first few days. I hardly moved out of my bunk. Thank heavens for the Aries and the trades. They kept us going in the right direction, and the other vessels that must have been around and avoided us.

Whatever the bug was, it worked its way through my system and after three days I started to take an interest in what was going on. I'd not kept the navigation up to date, and it was into the fourth day when I got some sights and established our position. I was still feeling off-colour when I arrived in Funchal and anchored in the harbour.

Madeira was a bit of a disappointment. There were other yachts there, but no yacht club or other focal point where we could meet. Landing for the dinghy was insecure, and you were advised not to leave the dinghy there for any length of time. There were shops ashore, and fresh food was plentiful. I daresay everyone is familiar little stubby bottles of 'Mateus Rose'. In Funchal I discovered numerous other brands Of Rose, all in similar stubby bottles. I sampled quite a few of them, but I wasn't there long enough to try them all.

I did a bus trip whilst we were in Funchal. It was just a regular service, but its route took us right round the island. I was away for something like three hours and when I came back the dinghy had disappeared. I could see some local kids fooling around in it and I started to yell at them. They waved back at me and kept fooling around. Eventually they headed for the shore about 200m away. They left the dinghy bumping against some rocks and scarpered. By the time that I got there they were out of sight. The dinghy was OK but was missing one oar and rowlock. I found it floating in the water nearby.

I'd seen enough of Funchal anyway, and I left the next

Chapter 4 - Circumnavigation

morning. It was only about 250nm to Tenerife our next destination, and the voyage was uneventful. The harbour at Santa Crus de Tenerife is large and commercial, but right up in the nor'-west corner is a small area set aside for yachts, and there we were all rafted up together, a little community of like-minded souls. One guy I remember clearly: a Scotsman from Orkney; a merchant seaman. He had a small, clinker-built timber yacht which he was taking to the West Indies for the winter season. He knew Tenerife well, and all the places that were worth going to.

I decided that I was going to do a bus tour round the island, and I went to a shop to book the tour. Come back at midday, I was told. When I got back there, I was told the bus has been cancelled, but there are a few people and we're taking you in a Mercedes. The other people were two husbands and wives, Argentinians. Things were a little stuffy to start with, but one of the wives spoke a little English and it was discovered that I was Australian, and I played a little rugby. They were thinking that I was English, but when they discovered that I was an Aussie they relaxed and out came the flagon of red wine.

It was a quite an enjoyable trip. The highlight for me was coming out of the clouds and into the sunshine as we neared the top. The road doesn't go right to the top, but we stopped and took the cable car for the last 200ft. The top, at 12,188ft, is the highest I've been on planet earth. It is also a volcano, and as I sit here labouring over the typewriter in 2022, I can tell you that one of the big stories of recent times has been the eruption of Pico Teide, and the lava flows running right down to the ocean.

When the Hiscock's visited Tenerife, they stopped at a place called Los Cristianos on the sou'-west corner of the island. It looked like a good jumping off place for the Atlantic crossing, and I thought I might stop and have a look. It is an open anchorage, but quite safe, as winds with a southerly component are very rare. I found the anchorage quite good with only a tiny bit of left-over swell working its way in. There is easy access for the dinghy via the beach, and shops with fresh food nearby. It seemed to be a bit of a tourist resort, and I heard a number of people speaking German. I had a pleasant enough few days there getting ready for the Atlantic crossing.

Tenerife is around 28 degrees north latitude, and my plan was to head south until we picked up the trades, then run down the longitude in the trade wind zone. However, most people that I spoke to left the Canaries and headed sou'-west, holding the northerlies most of the way across.

So, I gave us at least an extra 200nm, and we probably covered over 3000nm in getting to Barbados.

It was a slower trip to Barbados than I expected, and from about halfway across the breeze started to ease, and I had to hoist the spinnaker to keep us moving. One afternoon the sky started looking a bit iffy and looking like it might storm. Sure enough, there was a gust of wind which took the spinnaker aback, and wrapped it round the forestay, and I couldn't get it down. This happened in the late afternoon, and I had no choice but to climb the mast and untangle the mess. It was well after dark by the time I got it all undone and back on deck. It was badly damaged and was unusable at the moment. Perhaps it could be repaired; we'll have a closer look at it when we get to Barbados. In the meantime, get some sail on the forestay and let's get on with it.

It was a slow trip to Barbados: 32 days. The anchorage is open but quite safe off the town of Bridgetown. I was showing the 'Q' flag, but no one was coming near me. I'm gonna go ashore,

Chapter 4 - Circumnavigation

I thought, and see whether I can find them. It didn't take me long to find out that that day was a public holiday, and all services were closed. I found a shop that was open and got some fresh bread and eggs. I had a chat with the shopkeeper.

'Don't worry about them,' he said. 'They don't want to know you anyway.'

On the way back, I came to a pub. I'd noticed it on the way into town. Maybe I could have a beer or two. The bar was upstairs. There weren't many people around. I had a beer and got in a conversation with a couple of guys. We talked about cricket, what else! They shouted me a beer, and I shouted one back. All of a sudden, the place was getting busy. I looked around, all these black faces, all looking at me. I was out of place here. Time for me to go. I hadn't even officially entered the country yet, and here I was, gonna get into trouble for sure if I stayed around here. I shouted my friends another beer and made my apologies, glad to get out of the place in the end.

The next day, I met the people on a trimaran. They'd passed me on the day that I'd had a problem with the spinnaker. I'd been up the mast struggling to untangle it in the half dark when I'd seen them in the distance going past me. We had a chat about sails and so forth, and I told them the story about the spinnaker, and the fellow said his wife was quite good with the sewing machine, so maybe she could do something with it. I gave them the sail, and they gave it a completely new luff. The sail was now slightly smaller than it had been, but still useable, and would get plenty of use in the Pacific. They charged me $90, which was reasonable, and I was happy with that.

It was in Barbados that I came across West Indian rum. It came in flagons, which I was told was the equivalent of six normal bottles to the flagon. The price was astonishingly low, and the quality very high. It was smooth to drink, almost like malt whisky, and I built up a store of it which lasted all the way to Fiji. There were a number of different brands, all equally good. I can remember two of the brands, 'Cockspur' and 'Mt Gay', both of which hail from Barbados.

My plan was to be at the island of Bequia for Christmas. I'd been told it was the place to be. Bequia is 95nm almost due west of Barbados, and an easy 24-hour sail. The anchorage was very crowded, and we were right at the back, anchored in about 30' of water. On both sides of the anchorage were boom boxes blasting out rock'n'roll 24/7. There was no escape. I went ashore to reconnoitre, but I didn't see anybody that I knew, or wanted to know. There were some shops, and bars that looked and were very expensive. I wasn't going to spend any of my limited supply of money there and went back on board.

As I climbed on board, I noticed a yacht approaching. It looked like a charter vessel, and there were a group of young people on board, all American by the sound of it. They were trying to anchor. There was a fellow in the bow holding a small anchor, and another on the wheel shouting instructions. Drop it here, he yelled, and the guy on the bow let go the anchor with about 20' of light chain and some nylon line.

'Is it dug in yet?' the helmsman yelled.

It's not even touching the bottom yet, I'm thinking.

This went on for some time before they gave up and moved somewhere else. I didn't want you as neighbours, anyway, were my feelings. In fact, I wasn't very happy with Bequia at all and was looking at my options.

Just a half day sail away was the island of Mustique. Mustique had been in the news in recent times. Apparently, Princess Margaret and her latest boyfriend were shacked up

Chapter 4 - Circumnavigation

there. Well let's go over to Mustique and see if we can get invited in for a cup of tea, Ha! Ha! The only anchorage at Mustique was very rolly and uncomfortable, but I did go ashore, and had a walk around. I saw large expensive looking properties, but nobody around. No point in staying here. Get back on board, and head back to Bequia.

Too many visitors at Bequia. I went ashore to try and get some fresh food, but I could find only leftovers. I decided to leave Bequia next morning for St Vincent, a large island to the north.

It was only about 10nm to the southern part of the island. I wanted to anchor near Kingstown, the main town in the southern part of the island, and as I approached, I saw an area where yachts were anchored, near a short wharf. It seemed to be OK to anchor there, and there were dinghies tied up at the wharf. I went ashore and started walking towards the town. It was farther than I'd thought, but eventually I came to a general store, and they had everything that I needed.

By the time I walked back I'd been away for more than two hours, and as I reached the wharf, I could see that the dinghy wasn't there. Oh shit! There it was, some distance away, with a bunch of black kids cavorting around in it. I started shouting at them, and they were shouting back, and gesticulating. I had a problem. It might take some time to get the dinghy back. Luckily for me, someone else had seen what was going on.

A fellow in a launch with an outboard motor was going after them. As he closed in on them, the kids jumped into the water and started to swim ashore. He grabbed the painter and towed my dinghy back to me with the oars trailing in the water.

'Thanks very much for your help,' I called out to the guy. 'I don't know what I would have done without you.'

'They are bad kids around here,' he said. 'Are you thinking

of staying around?'

I said I just wanted some supplies; I'd be moving on in the morning.

'You could tie up at the wharf in town,' he said. 'They'll let you stay for a few days.'

I told him I might just do that. He waved his hand and revved the motor and was gone. The next morning, I got going early and went round to Kingstown and checked out the wharf. There were vessels at the wharf and no room for yachts. I carried on.

The next island in the chain was St Lucia. I didn't have any information about St Lucia, but there would surely be an anchorage for the night somewhere on the eastern side. It was less than 50nm from here and we should be there before dark. The breeze was very light, and we just kept motoring up the coast. We were obviously in a wind shadow. After an hour or so the nor'-easter started to come in and I immediately got the main and genoa up and away we went.

The breeze was increasing rapidly, and I had to reduce sail. I should have been ready for this. These islands are relatively high, and tend to block the wind, which funnels through the gaps between them. The 15knot trade wind increases to 25-30 knots and more at times. I'd read about this, of course, and should have been ready for it. We very quickly went from full sail to one reef in the main and #3 jib bashing into short steep seas.

We did the crossing fairly quickly and were up on the coast well before dark. We anchored off the town of Soufriere, and in the morning, we went up to the wharf and tied up. An official approached us and told us we could stay for a few hours only and had to be gone by late afternoon. I was getting the impression that many of these islands didn't really welcome visitors and would

Chapter 4 - Circumnavigation

rather be left alone. Anyway, I went for a walk around the town, and bought a few food items, and was off the wharf by the afternoon, and anchored where I'd been the previous night.

St Lucia is an independent country, previously a British colony, and the official language is English. Funnily enough, most of the names sound distinctly French, and the language spoken by the locals is a French patois.

Next morning, we hauled up the anchor and set sail for Martinique. The nor'easter came in early and was every bit as strong as the day before. We had a tough day's sailing and got to Martinique just before sundown. Martinique is French and is part of France, as the French do it. There is a different atmosphere there altogether and everyone is free to go there and enjoy their visit. At Fort de France, I anchored among numerous other yachts just off the main town and went ashore and saw 'les Douanes' and got a stamp in my passport, then I changed some money and visited the local markets, where I had some fun with my schoolboy French buying fresh fruit and vegetables, bread, and eggs. Most of the yachts were French, but there were at least two South African yachts, and yachts from other European countries, but no British yachts. I think it was either the third or fourth morning that I got up to have my early cup of coffee and went on deck to be regaled by a young lady on a nearby French yacht sticking her bare bum over the back of the yacht and doing a poo. Excuse me!!

There are several good anchorages on Martinique, and I visited all of them in the next week or so and enjoyed the peace and quiet.

Our next stop was to be Dominica, another independent island. The best anchorage looked to be at the town of Portsmouth on the nor'-west coast. It was a trip of over 50nm which meant we'd have to start early to get there before dark. It was a hard day's sailing; the wind was strong and gusty and never let up. In the end I had to tack to port to get us close to the shore just before dark. The anchorage was good, there were no other yachts around, and there didn't seem to be much activity ashore. Next morning, I launched the dinghy and rowed ashore and found a scene of desolation. I eventually met some people who told me that a strong hurricane had come through there eight months earlier and had devastated the place. I could see that the people were in survival mode, and they really didn't need people like me making demands on them. The best thing that I could do was to get the hell outta there. Dominica was only going to be one or day stop anyway, but it was another disappointing episode.

Back on board, I decided that I had to get on with it. I wouldn't stop at Guadeloupe; we'd go straight to Antigua. I definitely wanted to stop at Antigua for a rest, then go straight from there to Panama. I'd heard bad reports about the American Virgins, in particular, so my feeling was to forget about the Virgin Islands altogether.

Dominica to Antigua is approximately 90nm and would be an overnight trip. We must keep to the west of Guadeloupe, and it was likely that we would lose the breeze for a while. Not to worry; we could always hoist the steel spinnaker. We set off from Dominica in the morning, and straight away faced strong to gale headwinds, with much reduced sail. As the day progressed, the breeze did ease a little, and as we came up to Basse Terre, it started to ease right off. I was about to start up the auxiliary when a light westerly breeze came in—a sea breeze on the west coast of Guadeloupe. I hastily hoisted the genoa, and we cruised along the coast for the rest of the afternoon.

After we reached the northern end of Guadeloupe, the trade

Chapter 4 - Circumnavigation

wind reasserted itself. It was more moderate this time and never exceeded 20 knots. I had the full main and #2 genoa, and we were close-hauled. Antigua is set off slightly to the east of Guadeloupe and we couldn't quite lay the course. However, at dawn Antigua was in sight and it just required a fairly short time on the port tack for us to arrive at English Harbour, which is a safe and secure harbour, the best around, and the most secure even in the hurricane season. Everybody went there.

I knew it would be crowded, and it was, but that was where I wanted to be, and I got in there and squeezed us in. I wanted to be able to record that I had anchored in English Harbour and that was that. Antigua is another independent island but seems to have fared better than most of the others. They have accommodated the yachting fraternity and extensive tourist resorts and seem to do well from them. The British, in particular, seem to come to Antigua, especially the yachties. When you go ashore you find an area that has been developed where there are shops of various kinds, a bar and nightclub, and a Post Office.

I was in the Post Office when I noticed an incident worth relating. I just needed some stamps for postcards. There were quite a few people there and I was at the back of the queue. I could hear American accents, and one in particular, the loudest one, was obviously from the deep south. The guy behind the counter was a West Indian, of course; just trying to do his job. But loudmouth wasn't interested and was abusing this poor guy in the most vile, racial manner. It went on until an American woman in the middle of the queue started screaming at him. Loudmouth turned around and went for her, whereupon the whole place degenerated into a wild melee. Three or four of the guys grabbed loudmouth and dragged him out the door and threw him into the harbour. The (black) police then arrived and took control of the situation, and things settled down.

The poor guy behind the counter was in shock and totally stressed out. He couldn't continue, and the Post Office was closed for the rest of the day. We know these people exist, and there are lots of them in the USA, especially in the southern states. They mostly stay close to home where they can assert their bigotry at will and get away with it. This was a case of one looney who has decided to travel to little old Antigua and behave like he does at home and can't understand why people get upset with him.

Flag etiquette requires that I fly the national flag at the stern. It also allows me to show a small personal flag at the crosstrees, and I do that occasionally. The St Andrews cross is the national flag of Scotland, but I prefer to use the lion rampant, which is a Scottish flag from an older day and age. While I was at Antigua, I had the lion rampant at the crosstrees. One day, I'd been ashore and came back aboard and was on deck when I saw a launch approaching. It pulled up right alongside, and a voice with a Scottish accent said, 'Ahoy there! '

'Ahoy yirsel,' I replied. The guy was younger than me, but quite well put on, and spoke well.

'I'm from that big yacht over there,' he said. 'We noticed you were flying the lion rampant, and we thought we might invite you over for lunch. How about it, will you come and join us? '

'Thanks very much. I'd love to come over,' I replied.

The yacht had a large cabin, and I joined a group of young people. Someone gave me a beer and I settled in. They were English, and they were employees of a large country hotel in Cheshire. It was winter in the UK, and business was quiet, and the owner of the hotel had shouted them a trip to Antigua for a

Chapter 4 - Circumnavigation

couple of weeks, lucky people. Everybody was being nice to me, and I was feeling relaxed. There were a couple of nice-looking gals there, and yeah. Some other people were arriving.

'The boss is here,' I heard someone say.

Everybody turned to meet the new arrivals. One stood out, a very attractive woman of about my age. She was the boss; you just know it instinctively.

'Come over and meet George,' someone was saying. 'George, this is Christine… Christine, this is George.'

We sat down and chatted and chatted. Time went by, and most of the people had left. We went ashore and had a walk then stopped for coffee and chatted some more. What about dinner? I'm a good cook, she says. Come over to my place and bring a bottle of wine with you.

She owned a condominium not far from the harbour, and it was only a short walk. She was just a lovely person, and it was a privilege to know her. She had been married when she was younger to a wealthy businessman. He had died in a car accident, which made her a wealthy widow, and the hotel had been only one of many investments. Her staff loved her and would do anything for her.

We were together for five days before they all flew back to England. The last time we spoke, she said.

'Thank you… Can I ask a favour of you?'

'Sure. Just say,' I said.

'When you get to Tahiti,' she said, 'send me a postcard with a sexy, voluptuous gal on it, one that I can put in the bar.'

Done, and she was gone.

It took me a few days to come back down to earth and begin

to focus on my next move. The Virgin Islands, British and American, were out of the question. What could be there that would even come close to Antigua? No, it was time to move on, time to get back to sea, just doing the simple things. Antigua to Panama was around 1200nm. We would have the trade winds all the way and it should take us nine or 10 days. I used the spinnaker, mainly during the day, and we charged along. The breeze held steady at around 15 knots, sometimes gusting up to 20 knots. For a single hander this was sailing on the edge. Dorado averaged six knots for the trip, and we were in Colon in eight days with no help from any current—her best performance, to my knowledge.

The 'Aries', bless its soul, was magnificent. After I set it up, it required a minimum of attention from me. I didn't get a lot of sleep. Sometimes, I took the spinnaker down and had a good rest, but a lot of the time the adrenalin kept me going. The seas weren't big and there was no real swell, and the weather was warm and sunny. A very enjoyable passage.

I'd read all about Colon and the passage through the Panama Canal, and I knew what to do and what to expect, and how much everything cost. The money side of it was already in my budget, and that was all covered. So, when we arrived in Colon I knew where to go. I dropped anchor near the yacht club, launched the dinghy and went over to the club. They welcomed me straight away and gave me a membership for two weeks. They issued me with an ID card with strict instructions:

Keep it with you at all times; you won't get into the club without it. Then they told me to bring 'Dorado' into the creek, stern to the wharf, and anchor out front. That meant that I could step ashore, and in 12 paces I would be in the club. Handy!

Next day, I went to see the canal authorities about booking

Chapter 4 - Circumnavigation

our passage through the canal. They were helpful, but they had some requirements. 'Dorado' had to be inspected and measured. Mooring cleats were very important (some yachts had been refused passage through the canal because the mooring cleats weren't strong enough). On 'Dorado', the mooring cleats were stainless steel and welded to the deck. The inspector was impressed. The other thing was mooring lines. We needed four long nylon mooring lines. No problem, we had that plus. They also explained that there would be a pilot with us, and we had to have four deckhands, one for each of the mooring lines. They issued me with a certificate and told me that it would be 7-10 days before we went through, and I would get two days' notice.

The big problem for me was where I was going to get four deckhands. The canal was built, owned and run by the Americans, and they still owned and ran it when 'Dorado' passed through. American troops are stationed there, and many American nationals have spent their whole lives there. Zonies, they call themselves.

The yacht club was a special place. It opened in 1914, when the canal opened, and has remained open 24/7 ever since. It's a meeting place for all people connected to the canal, and the yachties who pass through find it a friendly and helpful place. The first time I went there, it was early evening, and the place was busy. I didn't know anybody, and I didn't stay long. I was still trying to catch up on sleep after the Caribbean crossing.

Sometime around 4am, I was woken by a noise, and I got up and stuck my head out the hatch to see what was going on. There was a cyclone wire fence behind us that ran some distance up the creek, and something was going on behind this fence. I could see the shape of a man running hard, running for his life down this fence. Not that far behind him were three or four guys in uniform.

They had guns and they were firing at the runner. Pop! Pop! Pop! went the guns, and I could hear bullets ricocheting around. I got my head down pretty smartly and waited until I could hear no more shots before venturing out again. I don't know whether they got him, but I certainly didn't feel like going back to bed. I was just a few steps from the club and of course there was the solution; maybe somebody in there will know what's going on.

There were a few people in the club and none of them seemed to be perturbed by what had just happened. Don't worry about it; we're all on the right side of the fence, seemed to be the attitude. Just an everyday event around here. There was a group of people over the other side of the bar. It looked like their night was over and they were going home. One of the girls didn't go with them. She came right round the bar and pulled up the stool next to me.

'Hello,' she said. 'I'm Penny... What's your name.'

George, I said. How did I get to be so lucky?

'You're not a local are you,' she said. 'You're a yachtsman, and you are shortly going through the canal, and you haven't got a crew yet. Is that right?'

'Yes, that's right,' I said. 'I've got a few days yet. I just don't know where to start looking.'

She said I shouldn't worry about it. It would be OK.

We chatted on for a while, and I discovered that Penny was a 'Zonie' born and bred. After school, she'd gone away to the mainland USA to study and become a nurse, then returned to Panama and worked in the local hospital. She'd been married but wasn't very happy (her husband was the master of an oil tanker. (The oil industry had built a fleet of tankers specially designed to fit through the canal and shipped oil from Alaska to the southern states of the USA for refining).

Chapter 4 - Circumnavigation

'I hardly ever see him,' she complained.

She went quiet after that. Maybe she'd gone too far.

'It's late,' she said. 'I have to go now, but I'll see you again. There's something I want to ask you about, OK?'

'It is late,' I said. 'Do you have a car?'

'Yeah, it's OK. The guy on the door will look after me. Bye,' and she was gone.

I thought about the fellow on the door, and I remembered what had happened the day before. I'd been out and was on my way back; I was approaching the club, and I could see the guard appeared to be talking to a rather scruffy looking bloke with long hair and a long beard. I got my pass out and showed it to the guard. He waved me through, but the scruffy looking bloke grabbed my arm.

'Excuse me, sir,' he said. 'This guy won't let me in the club.'

He said he was trying to contact someone; he told me a name.

'I'm trying to contact such and such, on such and such boat.,' he said. But before he could go any further, the guard began to wallop into him with a truncheon. Wow! I got myself into the club and watched as the guard pursued the bloke down the road. This place looked friendly and easy going on the surface, but there were definitely unpleasant undercurrents. Make sure you look after that member's pass, I thought, and don't lose it.

'Would you like another beer, sir?' the barman was saying.

'No, sorry mate,' I said, suddenly returning to the present. Shit! It's 5.30, almost dawn, time to get back to bed.

I didn't get back into the club until later in the evening. Penny was there with some male friends. I didn't approach them straight away, but Penny waved me over and introduced me. I sensed an undertone of aggression from the guys. They were suspicious of the interloper. Penny was talking about a launch that she had. It had a diesel engine, and she couldn't get it started. Would I come and have a look at it? Yeah, well, small diesel engines were pretty simple things really. There wasn't a lot that could go wrong with them. Sure, I'd come and have a look at it. I'll pick you up tomorrow morning about 9am, she said. No problem, see you then.

The launch was tied up at its own little jetty on the other side of town. I had a quick look over the engine. It was an American make; hardly surprising. There was fuel in the tank, the cooling system looked OK. There was an air filter with a cartridge inside. Took the cartridge out. The whole thing was wet and soggy and was probably blocking the inlet. Put it aside for the moment. It was a hand start; we'll give it a try. It was very hard to wind up, and when I dropped the valve lifter nothing happened. Try again… this time, Penny dropped the valve lifter. No joy. One more time… I really gave that handle a workout. Nothing. Time to try the secret weapon. I'd brought the 'Easi-Start' with me. Two scoots of 'Easi-Start' into the air intake, wind the starter handle up again, drop the valve lifter, and bang- tic-tic-tic-tic… It's off and running.

'Yahoodi!!! Thank you, thank you, George.' Penny grabs me and gives me a big kiss, and we stand there, embracing, in the tropical heat and humidity, the sweat running off us.

'You'd better come back to my place and have a shower,' she says.

The day changed course altogether, and we just enjoyed each other's company. Next morning, we drove up to the lakes above the locks, where the locals go swimming and dinghy sailing, and have barbecues. After lunch, she dropped me off at the club.

Chapter 4 - Circumnavigation

'I got things to do,' she said. 'I'll see you later.'

Another yacht had arrived at the club. It was the Australian boat, 'Australian Youth'. I'd heard about it, and the owner, single-hander Alan Kearney. He was from Adelaide, and he'd built his own steel boat and taken off around the world and, 15 years later, all sorts of dramas, disasters, illnesses later, here he was in Panama. I knocked on the hull but got no response. Not there, but there was another option. I walked into the bar. There weren't many people there, and he was fairly obvious drinking on his own. I walked over and introduced myself.

'Yeah, I was hoping to meet you,' he says.

The discussion continued into the evening, and suddenly it was late, and he was talking about going out on the town. I was smart enough not to go, but he was determined, and he went off with two other guys. I'd had enough and went to bed. In the morning I was feeling very second hand, and it wasn't till lunchtime that I went ashore to see what was happening. Alan wasn't around, but one of his mates from last night was there and came to me straight away.

'Have you heard?' he said. 'Alan is in jail. He played up last night, and tried to set fire to a brothel, and the Panamanian police have got him banged up. We're trying to get some money together to get him out.'

What money did I have? My budget was very tight, I had just about enough to get back to Sydney, and that was it. They were talking $3,000 to pay the coppers off. Well good luck to you mate, I can't help you right now. Sorry! I walked into the bar, and the barman came over to me straight away. He had a letter in his hand. You're going through on Friday morning, he said.

The letter was indeed from the Canal Authority and asked me to ring their number and confirm that we were ready. I wanted to see Penny first to make sure that I had a crew, and I asked the barman. She comes in most evenings, but later on, he said. I didn't want to get drunk again, so I went back on board. There was plenty to keep me occupied. After 9pm I strolled into the bar, and sure enough she was there with her workmates. I didn't approach them straight away, but after a while her friends left her, and she came over.

'I see you there,' she says. 'Have you got your letter yet?'

'Yeah, here it is.'

So, Friday morning, she says.

'Do you still want me to organise some crew for you?'

'Yes please… I've got nobody at the moment.'

'OK,' she said. 'I'll be here on Friday morning at 9am, with four young ladies.'

'Wow!' I said. 'That'll be fantastic. Thanks very much.'

She said we'd need food and drink.

'It'll be an all-day trip,' she said, 'and hot up there on the lakes.'

I asked Penny whether there were shops nearby where I could stock up.

'Don't worry about it,' she says. 'I know where to get the right stuff. I'll look after you.'

'At least let me pay for it,' I say, getting my wallet out.

'One of them will do just fine,' she says, plucking out a $50 note, 'and you'll all have a nice day.'

She had to work on Friday so she would not be one of the four, but she would see me on Friday morning to see me off.

Alan turned up the next day. His story was that the Americans got him out of jail after he paid the police chief $3,000, and they'd pushed him up the queue for the canal, and he was going through the day after me.

Chapter 4 - Circumnavigation

He asked me if I'd crew for him. I agreed.

'I'll come through on the train and be here by 9am,'I said.

I was suspicious of Alan's story. I felt that he might have exaggerated a bit. And what about the fellow who'd tried to get money from me? What was he up to? I'm glad I never gave him anything!

Friday came, and it all went off like clockwork. The pilot turned up, then Penny turned up with four young ladies. There is an International Women's Hostel in Colon and Penny knew the people who run it. Volunteers are not hard to come by when there is a trip through the canal on a yacht on offer, with food to go. Penny introduced them to me. One lass was from Norway, one from Germany, one was English, and the last one was a nurse from New York City. I barely got the chance to thank Penny before she left. One of the nicest people I've ever met.

So, pull the anchor up, and motor up to the lock. There is a ship there already, and we squeeze in behind it. Four monkey's fists come flying down; we attach our mooring lines, and they're pulled up. Now we secure our ends to their cleats. We're ready, and the massive door is closing behind us. Next, they flood the lock. The turbulence is powerful; that's why we need strong fittings. The lock is full, and we get ready to move forward. The ship goes first, pulled forward by a powerful winch, then after it is set, we move in behind it. Second lock is the same procedure.

When the second lock is full, we are then at the level of the lakes. The door at the front of the ship opens, and it moves forward into the lakes. The turbulence from its propellor is great, and we stay tied up until it is clear of the lock. Then it's our turn, and we come out of the lock and into the lakes. Now we can start on the food, and boy, did Penny do us proud. Beautiful fresh bread rolls with cold meats and salad. The pilot was envious as he munched his sandwiches. Soft drinks only though the canal. We have a job to do, and the canal authorities discourage alcoholfuelled parties.

The ship follows a dredged channel, whereas yachts can cut corners and follow a shorter route. The nor'-east trades extend across the isthmus, and sail can be used, especially across the lakes. It's expected that a yacht will go down in the same lock at Balboa as the ship that it came up through the locks with at Colon, so you've got to get a move on, no messing around. It is about 40nm through to Balboa, and that makes it a long day for a small yacht.

But it was an enjoyable day for us. The locks were negotiated with no problems; the sun shone, and the nor'-east breeze blew moderate/fresh. We actually stopped the engine and just sailed for half the distance and averaged nearly six knots for the whole trip.

The procedure for us after we came through the lock at Balboa was that we proceeded down the river to the Balboa yacht club where we were allocated a mooring. A service launch came and took us off, and into the yacht club where we all parted company. I thanked them all for their help and hoped they'd all had a nice day. They thanked me in return and went off to the nearby railway station.

The railway line was part of the canal setup, and trains ran on the hour between Colon and Balboa. It was free, and I would be using it next morning when I headed back to crew for Alan Kearney. I checked into the yacht club, where they gave free membership for a week (if I wanted to stay longer then it would cost me). I knew a week would more than cover me, so that was fine. I also discovered that the club was on an American military base, and there was a PX (store) just up the road, within walking

Chapter 4 - Circumnavigation

distance, which I was allowed to use. However, for official things like Customs and Immigration, I would have to take a taxi into Panama City. I didn't like the sound of that, but anyway, one thing at a time.

Just outside the club, there was a nice slipway, and nobody was using it. I enquired about cost, but it was 'members only', I was told. The lady then started to tell me about Taboga Island and the wharf that yachties used for scrubbing down and antifouling. I'd read about Taboga Island and had planned to go there anyway.

It had been a long day, and I had to get up early tomorrow, so I caught the service launch back to 'Dorado' and had a good night's sleep. Next day, I got the train over to Colon early and crewed on 'Australian Youth' as it transited the canal. After that, I sat in the club with Alan and had a few beers. He was going on to Costa Rica. He had American friends and was going to do some work for them. He said he would be back in Sydney before long and would be at the Cruising Yacht Club in Rushcutters Bay (he was a member of the Royal Adelaide Yacht Club and that gave him free access to all the top clubs). I promised to look him up and offered to take him to that well known pub, 'The Australian Youth Hotel', in Bay Street in Glebe.

Alan then got involved in a long phone call with some guy in the USA. It sounded as if he was trying to touch the guy for money. I found this a bit embarrassing and left him to it. Next morning, I came ashore and walked to the PX. It was very well stocked, especially with fresh foods, and so I filled up a couple of bags.

When I got back to the club, I decided to have a coffee, and got into conversation with a couple who'd been into Panama City trying to get a clearance for their yacht to sail to Tahiti. It involved going to two different offices in different parts of the city, and the taxi fares had totalled over $100. Balboa Yacht Club was in a very isolated position and there was no public transport, so if you didn't have your own car, you had to get a taxi. $100 was a lot of money to me at that stage. My bank was down below $500, I had enough basic stores to get me back to Sydney, and I would buy fresh stuff whenever I could. There was nothing left for emergencies or for playing up. I had developed a bad habit in Europe and the West Indies of not worrying about clearances in or out. If it was going to be that difficult, and expensive, then it had become easy to say, don't worry about it. A bad attitude that was going to bring me trouble for sure.

I spent another couple of days laying in stores and getting the boat ready for the Pacific. On the third day, I dropped the mooring and headed over to Taboga Island. The wharf was obvious on a nice curving beach. The tide was low, and the concrete block was also obvious, too. I anchored off the beach for the night, and in the morning, there was enough tide for us to go to the wharf and position ourselves over the concrete block. As the tide receded, I got busy with the hard scrubbing brush, and by low tide the hull was scrubbed back and ready for paint. Then it is a race against the tide to get the job finished. All the paint went on, four litres of the best anti-fouling, and that should be good enough to get us back to Sydney. A job well done. At high tide, we move away from the wharf and anchor just off the beach. Tomorrow morning, we'll be on our way.

Next morning, we haul up the anchor, and stow it away below deck.

We won't be needing it for a while. Panama-Nuka Hiva in

Chapter 4 - Circumnavigation

the Marquesas is around 4000nm, easily our longest leg. It will be light weather sailing. Every account I have read or heard tells the same story. There is a helpful current, maybe 25nm/day at best, but if we can do the trip in 40 days we will have done well. The nor'-east trade is blowing at 10-12 knots, so the spinnaker is up, and we're making five knots.

After two days, the breeze starts to fade. We're about to enter the doldrums. I change course, more towards the west, and put us on a broad reach. It is important to try and hold the breeze for as long as possible, even if it means going off course. 'Dorado' responds and our speed increases to four knots. Close to the coast, the area of the doldrums is quite wide. However, the farther west you go, the narrower the doldrums, and it's better to try and hold the breeze, and make some westing, even if it means a slightly longer trip.

We were about five days out, maybe about 200nm west of the Galapagos. I was below decks when I heard a noise. I came up on deck in a hurry to see what was going on. There was a helicopter approaching, making straight for us. It circled around. It carried no markings that I could see, but the door was open, and there was a man in the doorway in some sort of uniform. He was holding what looked like a sub-machine gun. I didn't like the look of that. What did they want? I didn't have a radio. I stood and watched them as they went around and around. They weren't American, I was fairly sure of that. They might be Ecuadorian, maybe from the Galapagos, but they were at extreme range. What could they want me for? They went round five times in all before heading back to the east. It struck me that it might have been handy if I'd had a radio. More or less, everybody has one these days and authorities expect you to communicate with them and are disappointed if you don't. I shouldn't have left

Sydney without one. A VHF at least!

Time was passing. We were more than two weeks into a six-week trip. We were still in the Northern Hemisphere heading west on a very light nor'-east trade. We were making reasonable progress, but we'd have to change course soon, and head sou'-west over the Equator, and on to Nuka Hiva. The days were pleasant, the seas were calm, and the spinnaker was pulling us along at four knots. One night, I'd been below on one of my two-hour rest periods. I came up on deck to check everything was OK. The sky was clear, but there was no moon. The stars were twinkling away as they do. City people never see the sky like this. A shooting star, then another, and another… Eh? Then I saw it, a large dark shape, shaped like a cigar… How close was it? Hard to say. Some miles away. All these little twinkling lights, some of them shooting out of it, others shooting into it and disappearing. What was all this about? I've never seen or heard of anything like this. The only light came from the shooting stars.

I stood and watched it for a while. The distance between us slowly increased, and the shape eventually disappeared. A mystery. I see something strange, and watch it for a long time, a strange feeling comes over me, I can't sleep, and get up and study the charts again. Then I go on deck and change course from west to west-sou'-west and readjust the sails accordingly. We're about six degrees north and I don't expect the trade to last for long.

The breeze held for two days, and then slowly faded and left us becalmed less than one degree north. No nasty surprises this time. No thunderstorms or wind squalls. I took the spinnaker down and stowed it. Only one thing to do now: start up the auxiliary and head straight south over the equator until we find the sou'-east trades. Leave the main up for stability. The Aries will still steer the boat as we are making our own light breeze.

Chapter 4 - Circumnavigation

It came about one degree south, a little puff of wind from the east, settling to a steady light breeze from the sou'-east. I hoisted the spinnaker again, and set the course for Nuka Hiva, just a little south of west. We're still around 2000nm from Nuka Hiva.

Time was going by, and I was starting to notice our progress wasn't as good as it should have been. One day I was looking over the stern and I noticed that the paddle of the Aries gear had a few gooseneck barnacles on it. Oh no! Not the dreaded gooseneck barnacles. I'm going to have to get in the water and check the hull. I tie a safety line around me and in I

go. A brief glance is enough; the whole of the hull below the waterline and the keel is carpeted with them. No wonder we're slowing down.

Back on deck, I work out a plan of action. I'm going to try and get rid of at least some of them. I tie a strong line from a stanchion on the port side and loop it down under the keel and back up to a stanchion on the starboard side. Taking my trusty scraper with me and using the strong line I pull myself under the water and start work. Gooseneck barnacles are a marine pest that exist mainly in tropical waters. They look a bit like a small thumb and are not deterred by anti-fouling. A sharp scraper chops them off pretty quickly, but I can't hold my breath for long, so it is hard work.

I'd been going for maybe 30 minutes when something caught my eye. A grey shape cruising around in the water not far away. Shit! How long have you been there, I thought as I shot back on deck? The shark, about three metres, cruised around close to the boat, then went to the stern and checked out the Aries paddle which was waggling around. It came right up behind the boat, as if it is sniffing it out. Then it turns around and cruises off… nothing to see here. I wasn't game to go back in the water; it'll have to wait until Nuka Hiva, was my thinking.

There was still a long way to go to Nuka Hiva, not far off 2000nm 'Dorado' was making less than three knots through the water, but navigation was showing that we were getting some help from the current, and our daily runs were not far off 100nm — not great, but we can live with that.

And so, we continued day after day. The weather remained constant, and we enjoyed an apparently endless spell of stable conditions. (To be fair, every account I've read or heard of the people doing this crossing have experienced similar conditions). Then one morning there is a small dot on the horizon almost dead ahead. It is Ua Huka. We expected to see it, but after such a long time at sea it is hardly believable. Soon, Nuka Hiva starts to appear, and we can believe it.

We pass to the north of Ua Huka, and the presence of land changes the strength and direction of the wind. The spinnaker needs management, but we hang on. Soon we're off the south coast of Nuka Hiva, and I can see the bay, and the yachts anchored there. Drop the spinnaker; it's done its job. Where would we have been without it! We motor into the bay and drop anchor. One giant step on the way back to Sydney.

It's late in the day and I'm not going ashore tonight. I have a few sips of my Barbadian rum, fantastic. I still have two flagons left. Will that get me back to Sydney? I doubt it.

Next morning, I have to see the authorities. My friends on the other yachts direct me where to go. He's a policeman, they tell me, a badtempered policeman, watch out for him. I don't tell them that I haven't got a clearance from Panama. I go ashore and walk up to the house, a police station. He's there in his office.

'Bonjour, bonjour, sir. I have just arrived from Panama.'

Chapter 4 - Circumnavigation

I show him my passport and ship's papers. He looks at what I've given him, and then his head comes up, and he starts yelling at me in French. I don't have to be able to speak French to understand what he is saying. No stamp in the passport, no clearance papers from Panama, you cannot stop here. You must go straight to Tahiti.

'Monsieur,' I say, 'the bottom of the hull is covered in gooseneck barnacles. I must clean them off before I go anywhere. And I have some sail repairs that need to be done before I go.'

He stares at me with eyes that look like they're going to pop out of his head.

'Deux jours,' he shouts at me, showing me two fingers, 'Deux jours.'

I feel like I've been assaulted as I walk away from his office. It's my own fault, of course, so I've got to live with it for now anyway.

My friends on the other yachts had told there was a bakery and where to find it. It wasn't very far, and I noticed the beautiful smell before I got there. It has been six weeks since I've tasted bread like that. It was a special treat, irrespective of other problems.

In the afternoon, my friends came to see me, and I told them what had happened. Don't worry about him they said, he's a nasty little man. He has treated other people the same way. They said they were going for a walk the next day, to the top of the hill (Nuka Hiva is 1200m high).

'There is a road all the way, they said. 'Would you like to come with us?' Yeah, why not. I'd love to. It is a dirt road all the way to the top and heavy vegetation lines the road both sides. I haven't had a walk like that for a long time. There were trees laden with fruit, and bushes, just help yourself, and a spectacular view from the top. It took most of the day, and I certainly didn't feel like doing anything else when we got back.

Next day I got stuck into the hull, it was exhausting work, dragging myself under water right down to the bottom of the keel, holding my breath for as long as I could, and getting as much done as I could before surfacing, and gasping for air. I didn't see any sharks, but there were lots of saltwater piranhas which come in shoals and surround you and nibble at your skin. Each nibble gives a little tingle, and that is what I lived with all day.

By late in the afternoon, I was not far off being finished, but it was getting to that time of day when sharks start to feed, and it was wise to get out of the water. I could finish off the bit that was left in a couple of hours tomorrow morning. It had been a tough day and I didn't have any trouble sleeping that night.

In the morning I went ashore and got some fresh bread and eggs, then came back and finished off the hull. That was a relief. The boat would sail much better now. I was sitting in the cockpit with a cup of coffee relaxing when I heard an outboard motor coming our way. I got up to have a look, and sure enough it was the policeman, and he was pulling up alongside us. He started screaming at me in French. My schoolboy French wasn't in the race, but I didn't have to be told what he was saying. He'd given me two days to get the hull done, and we were into the third day. Leave now, he was screaming, go straight to Papeete, 'tout de suite', don't stop anywhere else. Something else he was saying about coming back, I could only guess at. Then he left and sped off round the point. I started to tidy up and get ready to hoist some sail, when my friends came over to commiserate with me. They said I wasn't the only one; there had been two other American boats in the last few weeks that he had chased away,

Chapter 4 - Circumnavigation

miserable person that he was.

I wasn't worried about leaving. How long would I have stayed anyway? A few more days. Nuka Hiva is a beautiful, peaceful place, and I could enjoy that too, but I was on a schedule to be back in Sydney in September/October, and that was what was guiding me.

Nuka Hiva to Papeete is just under 800nm, and we will be in the trades for the whole trip. A piece of cake!

Not quite. We have to pass through the Tuamotus on the way. The Tuamotus are a string of coral reefs which extend right across the path of vessels heading for Papeete, and the sailing directions says this about them: It is best to navigate around the Tuamotus during the day. The area is unlit, and can be subject to local weather conditions, and strong tidal currents. So, we'd been warned.

On leaving Nuka Hiva, the breeze was quite light, and I set the full main and genoa. The course was sou'-west, and if we held that course it would take us between Aratula reef and Rangiora reef, and then on to Papeete. The gap between these two reefs was the biggest I could see, and it seemed the sensible way to go. By the fourth afternoon, we were about 60nm from the pass. We'd be there about dawn.

It was an anxious time. I couldn't allow myself to sleep. As we approached dawn I was on deck, ready, a faint sound, breakers on a reef, the noise started to increase, as the dawn started to break... I could see palm trees... we're in trouble! I eased the sheets, still going the wrong way. I started up the auxiliary. No messing around now. We were being set down by a current. We'd missed the passage by 1nm, and now we had to fight our way back. It took some time to get to a position where we were out of danger, and we could resume our course, and settle down. It was a sharp lesson, wasn't it. It's so easy to get into trouble, even when you're ready for it.

It took us another two days to get to Papeete. We arrived in the afternoon and took a position 'sur la plage', as they say. There were plenty of yachts there; plenty of friends to make. One of them was Dave Benson, an Irish/English/Australian, who'd built his own steel yacht (a 38-foot Alan Payne design) and sailed it single-handed (no self-steering gear) all the way through the Pacific to the Panama Canal, then all the way up the Atlantic to Ireland. He spent the winter in England and came back the same way. A rather unusual way to do it. However, each to his own. I'd heard about him in the West Indies, and just missed him at Panama (he'd gone up to Costa Rica looking for work), now here he was in Papeete.

We became friends straight away and had a couple of nights out. This was limited by the astronomical price of alcohol. Beer is brewed in Papeete, but in spite of that a beer cost five or six times the cost in Sydney. We went twice to restaurants, but our lack of French was a problem both times. As I walked along the beach, looking at all the yachts, I see a red ketch of about 40ft. It is 'Joshua', and there is the man himself, Bernard Moitessier, surrounded by friends. I would love to have had a one on one with Bernard, but it was not to be. I saw him two or three times, each time with other people around him. Bernard is a bit of a hero to me, a first-class seaman (Joshua does not have an auxiliary), his books 'The Long Way' and 'Cape Horn the Logical Route' are very special to me. He seems to have a very spiritual connection with the sea, something that I feel myself at times.

One thing I wanted to do in Papeete was to visit the spot

Chapter 4 - Circumnavigation

where Captain Cook observed the transit of Venus. In fact, it was one of his major tasks. Nowadays there is a small museum at Venus point, and I wanted to see it. I thought I could walk there, but I was told, no, it was too far to walk. Take the bus. They run every few minutes. So, I took the bus and the driver put me down at the right stop. There wasn't a lot to see at the museum, but it was good to stand on the same spot where the famous man had stood to make his observations. Astronomy, and navigation were still at a basic stage in those days. Navigators could calculate latitude but had difficulty with longitude. For safety, ships relied on a man in the crow's nest 24/7.

It was late afternoon when I got back to the beach. When I went to get back on board, one of my neighbours called out to me. The Customs were down here this afternoon looking for you, they told me. They want you to see them straight away. I should have gone and seen them, shouldn't I. Now I'll be in even more trouble. Well, it's too late in the afternoon now. I'll worry about that in the morning.

In the morning, I got up early and pulled up the anchor, motored out through the pass and went over to Moorea! I did what? What was going on in my head that morning? As I write this down 42 years later, I feel embarrassed about it. I should have gone to the Customs. The worst that would have happened would have been a ticking off, and perhaps they would have asked me to leave, which I was about to do anyway. I had started something that was going to follow me across the Pacific.

Moorea is not far from Tahiti, about 20nm to Cooks Bay, where I had decided that I was going to anchor. I was there by lunchtime, after negotiating the extensive reefs to the north of the island. I had a chart of the island, and that was very useful. I anchored close to the road on the eastern side of the bay, clear of an area marked as foul ground. It was a very beautiful and peaceful place; I had it all to myself. There was a road which runs around the bay—around the entire island of Moorea—and there was traffic, of course. I'll go ashore tomorrow, I thought, and walk down the road. It would be nice if I could find a shop.

I walked for over 4km, and I did find a shop, and bought some fresh supplies—just what I needed. On the way back, I'd gone only about a kilometre, when a vehicle came from behind me. It pulled over and a couple of guys got out.

'Would you like a lift,' one of them said. 'How far are you going?'

'Not very far,' I said, 'but thanks all the same.'

'We were wondering if you had any guns for sale,' the fellow said.

'We'll buy them from you?'

I told them I didn't have any guns.

'OK,' he answered, and they got back in their vehicle and took off.

Just near where they stopped, there was a little shop by the side of the road. I'd noticed it on the way in, local artefacts and knickknacks. There was a guy standing outside it.

'Hi there,' he said. 'I'm Bill.' The accent was American.

'I couldn't help noticing those guys stopped and spoke to you,' he says.

'Were they asking you if you had any guns for sale?'

'Yeah, they did,' I replied. 'They offered me a lift, and then took off when I told them I had no guns.'

He laughed.

'There are some funny people live around here,' he said. 'You should be careful.'

We chatted for a while, and I found that he'd come to Tahiti

Chapter 4 - Circumnavigation

during the Viet Nam war when he was about to be called up for military service.

'I was like a lot of them,' he said. 'No passport, no paperwork of any kind... We survived by selling guns, drugs, that kind of thing.

'A lot of people had little plots of drugs up there in the bush.

'We caused a lot of trouble in those days. The American Embassy was swamped by destitute people. After the war ended most of them went back, but some stayed, and scratched a living doing this and that.' 'So, this is your this and that,' I said, looking at the artefacts.

He laughed. 'Yeah, I don't do too badly. I survive. I kinda like it here. my folks at home have disowned me. I disgraced the family by running away, and they won't have me back.

'Just be careful,' he said as I walked away. 'Don't leave your yacht unattended for any length of time and take your dinghy on board every night.'

It was dark when I got back to 'Dorado', and I was glad to be back. I took the dinghy back on board, and lashed it down, just in case.

The next day, I went ashore again and walked in the other direction up towards the end of the bay. There wasn't a lot to see but I did find a shop and managed to get some fresh stuff, which was good as I was planning to head out the next morning. Moorea was a beautiful place, but I had things to do.

Rarotonga was going to be my next stop, about 650nm away to the sou'west. At 21 degrees south, Rarotonga was still technically in the trade wind zone, but this was winter, and the westerlies occasionally reach that far. I slipped away early in the morning and was out of sight of Moorea by midday.

The trades held for nearly three days, then faded out. The barometer was falling, and a light breeze came in from the nor'-west. It quickly freshened up and shifted to the west. It got up to about 25 knots for an hour or two, then shifted to the sou'-west at 30 knots. The barometer started to rise, but the wind stayed sou'-west. We hadn't done any sailing like this for quite some time. One reef in the main, and #3 jib, bashing into short seas with spray flying everywhere.

We weren't able to lay the course for Rarotonga, but I held onto the starboard tack hoping the breeze would shift to the south, which it did after a few hours, and we were able to tack back to port and lay the necessary course. It also began to ease a little, and by the next morning it had eased to sou'-east 15-20 knots. Not long after dawn, Rarotonga was in sight, and we were able to enter the harbour before dark.

The harbour is very small, and even the few yachts that were there made it crowded. Yachts were expected to drop anchor in the harbour and then manoeuvre backwards into a small quay. We are all effectively rafted up, and good fenders are required. It is completely open to the north, and there would be chaos if even a fresh northerly came along.

Next morning, I went along to the Customs office to explain myself. They were surprisingly nice about it. They put a stamp in my passport but wouldn't give me a clearance other than that. They told me, 'Look, next weekend is a holiday weekend, and we want you to leave by

Friday, OK?'

That was six days away, and probably as long as I would have wanted to stay anyway. In the meantime, I got to meet some

Chapter 4 - Circumnavigation

new friends, and had a pleasant few days. Rarotonga is the main island in the Cook Islands group, and some of the yachts were going on to other islands. But I was quite happy where I was. (The Cook Islands are self-governing but in free association with New Zealand.)

Rarotonga and Nuku'alofa, the capital of Tonga, are on more or less the same latitude, and the separation is close to 900nm. The voyage was uneventful and took under eight days. The trades were moderate to fresh, more sou'-sou'east than sou'-east, and 'Dorado' was on a broad reach, which she really likes. We arrived off Nuku'alofa in the late afternoon, and were able to negotiate the tricky entrance channel, and get into the small boat harbour before dark. It was crowded but we managed to squeeze in.

I went to visit the harbour authorities in the morning. They seemed to know all about me. They stamped my passport and asked me how long I wanted to stay. Seven-10 days I said, but then I'd like to go up to the Vava'u group in northern Tonga. OK, they said, but I must see them before I went. No problem.

Tonga was a friendly place, and I enjoyed my time there. I was starting to feel closer to home, and that was reinforced when I walked up the street and saw taxis all decked out in the Red Deluxe colours that I knew so well, and they even had radio numbers on them, numbers that were familiar to me because I'd driven these very taxis. How could that be? I discovered later that Red Deluxe had replaced a bunch of old taxis and sold them in a job lot to Tonga. And here they were running up and down the streets of Nuku'alofa.

There was a social club not far from the boat harbour, and yachties were welcome. I was in there one day when I met a fellow named 'Jocka'. His accent was broad Glasgow, which explained the name. He lived at the royal palace, he said, and was employed by the Tongan royal family. He was an expert at tricky little jobs like building model planes etc, etc, and made himself indispensable to the family, and was much loved by all, including the princesses, he told me.

'Do you think I could come out and meet some of the princesses?' I asked him.

'Sorry, no chance,' he answered. If I wanted a gal, he advised me, there was a building not far from the social club, a three-storey building, the only one in Tonga.

'Go up to the top floor, and you will meet some nice gals there,' he said.

I took him at his word one day and went and found the three-storey building and went up the stairs to the top floor. There was nothing and nobody there. I hung around for a few minutes, and two young gals appeared. I tried to approach them and have a conversation with them, but they just started giggling. The harder I tried, the harder they giggled. I soon got fed up with it and walked away and left them. A few days later I was shopping at the Burns-Philp store in town when suddenly I heard an outburst of giggling behind me. I turned around and there they were, and everybody looking to see what was going on.

My last attempt at some social life was to visit the pub. The International

Dateline Hotel was built at the exact spot where the International Dateline crosses Nuku'alofa. I hadn't heard much about it but thought I might investigate. My financial position was not good, but maybe just a couple of beers. So, I walked in and fronted up to the bar. I ordered a beer and glanced around me. The place was full of Islander queers. I can't believe it; I'm getting come on glances right left and centre. I grab my beer and

Chapter 4 - Circumnavigation

swallow it down and bolt. Bloody Hell, even in a faraway place like Tonga.

The islands that make up the Tonga group stretch away to the nor-noreast and Nuku'alofa to Vava'u is approximately 230nm. For the navigator, there are many low-lying reefs, tidal currents, and also a magnetic anomaly to contend with. There is also a lighthouse on the south side of Vava'u with a limited range. The volcanic island of Tofua lies about halfway and would be a good mark for me. I was expecting that it would take me a good two days. I left the boat harbour early one morning, but the breeze was light, and progress was slow. It was the next day before we passed Tofua. There was smoke coming from it, and I could see a red glow.

Progress continued to be slow. After dark, everything seemed to close in around us, but the sky was clear, and I was confident that I would be able to see land if there was any to see. In the early hours of the next morning, I started to hear surf breaking on a reef, and it seemed to be getting closer. We should have been seeing the light on Vava'u by now, and I was starting to become concerned. The safest thing would be to heave to and wait for dawn.

Four anxious hours passed, and I continued to hear surf breaking till dawn arrived and solved the puzzle. Vava'u was straight ahead of us 7-8nm away, and there was the light that I was expecting to see. It was out. There were reefs to the east of us at a safe distance, and no other dangers. We proceeded and went right up to the base from where the charter boats operate and dropped anchor, launched the dinghy, and went ashore. The people at the charter yacht base were quite happy to see me.

They said they would report me back to Nuku'alofa and gave me a map of Vava'u marked with all their anchorages. They told me the locals would probably approach me and try to sell me their produce.

'That's OK,' I said. 'I love bananas.'

Just as well, I ate bananas all the way to Fiji. I also met and had conversations with some of the charterers. It was funny meeting people, who a few days earlier had been in their homes and offices in Sydney, Melbourne, Auckland, etc, and were trying to adjust to this faraway place, and its marine environment. It wouldn't be too long before I'd be back among them.

We stayed in Vava'u for over a week and covered all the best anchorages. Time was moving on, and I was planning our next move. The direct route to Fiji meant navigating among a series of reefs which were unlit and were surrounded by strong tidal currents. I didn't like the look of that too much and decided to avoid the area. At the southern end of this area, there were two continental islands, Ongea and Totoya, which would be marks for us to round, then head straight for Suva. The extra distance is not great, no more than 50nm, but I figured it would be a much safer route.

I went back to the charter boat base and told them what I was going to do. We can't give you clearance here, they told me. I'd have to go back to Nuku'alofa for that.

'Oh, right oh. I hear what you say,' I said. 'No problem. Bye now.'

I thought I'd cleared all that up before I left Nuku'alofa, and I certainly wasn't going back there now. Just another sad chapter in my dealings with officialdom. Let's see what Fiji brings.

The voyage went well. No problems. The trades were moderate/fresh, and we were in Fiji in under six days. Technically, I completed my single-handed circumnavigation of the world when I arrived in Suva, but I wanted Suva to be low

Chapter 4 - Circumnavigation

key. Getting back to Sydney would be the big celebration, and right now my bank was just about gone. So just a few quiet days in Suva to rest up would be fine.

We came through the break in the reef into the calm of Suva harbour, headed over to the quarantine area, dropped the anchor, and set the quarantine flag, and waited, and waited. I was just about to pick up the anchor when I saw a launch leaving the wharf and coming my way. They came right over and tied up alongside of us. There were two Fijian Indians on board, and I didn't like the look of them as they boarded us. You should have called us on the radio, they said, and then come into the wharf. I told them I didn't have a radio, and I was just doing the same as I had the last time I was there.

'You should have come to the wharf,' one of them said. 'Now we've had to work overtime to come and see you, and we'll have to charge you for that.'

This pair were getting on my wick. I'm at the end of a circumnavigation, and not one port that I have stopped at has charged me as much as a cent for practique, I said, 'and I won't be starting with you'. They started getting back on board the launch.

'You won't be leaving here until you pay for our overtime,' they shouted at me, not quite in unison. Well, I've stuffed up again. I definitely should have installed that VHF radio. Now I've got this awkward situation again. What to do? The obvious thing. Go over to the yacht club. I can see a number of yachts anchored over there. We anchored at the back of them, not all that far from the mangroves, but there was enough water. We should be OK. Launch the dinghy, and row over to the yacht club, I spoke to the barman.

'The office is closed tonight,' he said. 'Come and see them in the morning.

'Have a beer,' he says. 'We always welcome a newcomer with a beer.'

That was a start. Before long, I'd made some friends, and was starting to settle in. Sometime later, well after dark, there was a commotion; something had happened. People were leaving. A heavy barge had broken loose from the wharf and was bearing down on the anchored yachts. Wow! I rush outside. It's pouring with rain, and the wind has gotten up something shocking; it's blowing a gale. I could see the big barge; it was about halfway across the bay heading straight for the anchored yachts. I got in the dinghy which was already half full of water and rowed like mad back to 'Dorado', pulled the dinghy on board, emptied the water out of it, and stowed it away. Then I started the auxiliary. Now we're ready to move if need be.

The barge was approaching the yachts, but there were at least three guys out there in large dinghies with good outboards, and they were having some success. Slowly the barge changed course. From where I was at the back, it got mighty close to the front row of yachts and couldn't have missed them by much. It took about an hour, but they got the better of it, and pushed it sideways until it was clear of the yachts then just let it go and it disappeared into the mangroves. Phew! You could almost hear the sigh of relief from all the yachties.

They were getting back into their dinghies and heading back to the club. It was later than we thought, though, and the club was just closing for the night. One of the guys who had been in the forefront of saving us all from the barge said, 'Come on out to mine… We have a big cabin and plenty of room.' So, all the dinghies converged on his yacht, which was the largest one there, and people crowded into the main cabin. People were

Chapter 4 - Circumnavigation

introducing themselves and sitting down at the main table. A lady came and sat next to me. I'm Laura she said, and this is my husband, Hal. We're schoolteachers and were from Texas.

'Hi. I'm George, and I'm a Sydney taxi driver.'

Before long, everybody had had a drink or two and the place was humming. Laura was telling all and sundry about how their boat's engine had broken down, and how Hal was trying to fix it, and how there were dirty bits and pieces lying around everywhere.

'I've had enough of it,' she said. 'I'm going for a run tomorrow.'

I'm thinking, what is she saying? She's going for a run in this weather. We were fairly tightly packed in there anyway, but I could feel her leg rubbing against mine, almost deliberately. In a low voice she said right in my ear.

'What about you, George? Would you like to go for a run tomorrow?'

'Yeah, I hear what you say,' I replied. I wasn't sure I knew what she meant, and she was probably talking through drink anyway.

We were starting to outstay our welcome. We thanked the guy for his hospitality, and for saving us all from the barge. Outside, the wind was howling, and the rain was still pouring down. Thankfully, when I'd gone back to 'Dorado' earlier, I'd taken off my wet clothes, and put on dry things, and my wet weather jacket over the top. It had been a long day, and I was exhausted. We were lucky that this weather hadn't caught us at sea.

I slept well. When I got up next morning, I was feeling refreshed. It was still raining, and the wind was still blowing a gale. I couldn't stay on 'Dorado' all day. I'd go over to the club, I thought. I still haven't checked in with them. Also, I was out of rum, and I'd been told they sold cheap rum there. I'll get enough to get me back to Sydney. Get the wet weather gear on, get in the dinghy and row ashore to the club. Tie up the dinghy to the pontoon—only dinghy there, not many people around the club today—walk into the club… Nobody there. No, there's one person there; it's Laura from last night. She comes over to me.

'George, you came,' she says, touching my arm. 'I was hoping you would. Shall we go and have a run together?'

I guess a smile must have crossed my face because she grabbed my arm, 'I've got a taxi waiting, and we're going to the best hotel in town.'

Mentally, I'm thinking but, but, but… but by the time we got to the hotel

I was warming to the idea. It was the next day before I got back to 'Dorado'. We'd 'run' ourselves into the ground and needed to reset. Sleep and rest were the orders for that day, and I didn't surface until the evening. I wondered whether it would be safe to go ashore. The need for alcohol prevailed, and I went to the club, and had a chat with a few people. I brought up my problem with the Customs officials, and the advice was unanimous. You're not alone; they are a set of bastards. Lots of people have trouble with them. Don't worry about them, was the consensus; when you are ready to go, just go. That's what I'd done in Tahiti, wasn't it, but I couldn't go direct to Sydney and arrive there with no paperwork. Customs were hard liners in Sydney, and I had form, as they saw it.

The answer was, go to New Caledonia first, and at least get my passport stamped there. That was the plan of action, and as

Chapter 4 - Circumnavigation

soon as this weather eased, I was out of Suva. The barometer had risen, the wind was still strong, but not gale force, and the rain had eased to showers with sunny breaks. Tomorrow was going to be the day, and I was going to leave early. I needed to do some shopping, mainly for fresh stuff, and 'Dorado' had to be made ready.

By lunchtime I was ready to go ashore. Getting some Anglo style fresh stuff wasn't as easy as I thought, and it was late in the afternoon before I got back to the club. Time for a beer before I get back on board. Some of my yachtie friends were there, and I joined them. I was probably on my third beer when I got a tap on the shoulder. It was Laura.

'How are you?' she says. 'I thought you would have been gone by now.'

'Yeah, tomorrow morning, I told her.

'I just want to introduce you to Hugo,' she said, and the big guy standing behind her stuck his hand out. He was clean shaven and looked a bit Spanish to me. Have a nice day, I said, as they walked away. I still hadn't gotten my bottle of rum, so I got that, and headed back to 'Dorado'.

I woke up before dawn, got up and had a coffee, then into it. Started up the auxiliary then hauled up the anchor, and slowly moved away from the other yachts. There was just a light breeze on the harbour as we motored across it towards the pass through the reef. The sun was just coming up as we passed through, and straight away the breeze started to increase. I set the full main and the #2 genoa, set up the Aries selfsteering gear, trimmed the sails, and set the course for New Caledonia, put the first entry in the log, turned off the auxiliary, and away we went.

The breeze was sou'-east force 4-5 the sun was in the sky. The gale was gone, and it was a pleasure to be back at sea. It had been five years since I'd last done this trip. I'd had no Aries then and spent long hours on the tiller. My navigation was under stress, and I didn't have the confidence to use the passage through the reef at the sou'-east corner of New Caledonia. Instead, we had gone right down around the reef, and had a couple of near misses before entering at the Amédée pass. This time, we went straight to the sou'-east pass and straight in following the leads and channel markers on a rising tide. Turn to the west and the land soon opens up into a large, secure bay. Good enough, we think, anchor here for the night, and head up into Noumea tomorrow.

Next day, we started early but the breeze lets us down, and we end up motoring most of the way to Noumea. The previous two times we'd been here, we'd gone to the yacht club, but this time I decided to anchor in Moselle Bay, where there was a park on the starboard side, and access via a set of steps. It's quite close to the Customs office, just a short walk. By the time we arrive, it is too late to visit 'Les Douanes'. That's for tomorrow. In the morning, I thought it through. It had been five years since my last visit. Would 'Maigret' still be around? I felt he was my best chance of getting out of this.

It was with trepidation that I walked along the wharf towards the Customs office. There were two people in the office, the young one rose straight away, then the older one noticed me and came over to serve me. It was 'Maigret'. How lucky can you be! He listened to my story patiently, then he said.

'M'sieur, you have come here from Fiji with no Customs clearance, and no stamp in your passport,' he said. 'I cannot give you Customs clearance, but I will stamp your passport.

'When you leave here to go back to Sydney, make sure you come and see me first.'

Chapter 4 - Circumnavigation

Well, that was as good as I could have expected; it was a step in the right direction. My money was running out, but I could at least have a few days in Noumea. Actually, I stayed for another 10 days. It was a time for reflection, and to think of the future. I'd enjoyed the last two and a half years, there was no doubt about it, and I wanted to continue sailing, not the same thing again, but a new challenge, the Southern Ocean, and Cape Horn. That would be an achievement worth trying for. I'd do it in 'Dorado', the boat that I knew. I'd make a few alterations to the keel to help with the weather helm and go over every little bit of it very carefully. The 'Aries' was in good shape; it had the reputation of lasting forever—just keep the moving parts lubricated!

When I went back to the Customs for my clearance to leave for Sydney, I got a good talking to. Apparently, the Fijian Customs had been telling everybody to watch out for 'Dorado'; that I owed them money. I could expect to hear about it when I got back to Sydney. Then they gave me my clearance for Sydney and stamped my passport. That was important. Sydney couldn't say too much to me if I had my papers from my previous port.

My third voyage from Noumea to Sydney was a lot easier than the second. We left New Caledonia with a fresh trade wind, which slowly petered out as we approached the Australian coast, from then on, we had westerlies, not too strong, all the way down the Australian coast. It was a bushfire year in eastern Australia, and I had the smell of smoke in my nostrils.

The smoke got thicker as we went south, and navigation started to be a problem. I was hoping for a southerly change to clear the air so I could see where we were. A change was forecast, and it arrived at just the right time. I reckoned I was somewhere off Sydney's northern beaches, and sure enough the breeze arrived and cleared the air, and there we were, sitting off Long

Reef.

It was an emotional afternoon as we beat our way around North Head then eased sheets as we cruised down the harbour to Neutral Bay to see the Customs. My visit to Customs was just a formality. Fiji was mentioned and passed off with a laugh. They congratulated me on completing my circumnavigation, stamped my passport, wished me well, and said, 'All done, sir. You're free to go.'

Just a footnote to this part of the story. During the two and a half years that I was away, I did not shave, nor did I have a haircut. My hair was down on my chest, and way down my back. I got several negative comments when I got back. Somebody took a photo of me and when I saw it, I cut the lot off straight away. It may well have been responsible for at least some of the problems that I had with officialdom in the Pacific.

GEORGE

GEORGE & CHLOE

DORADO

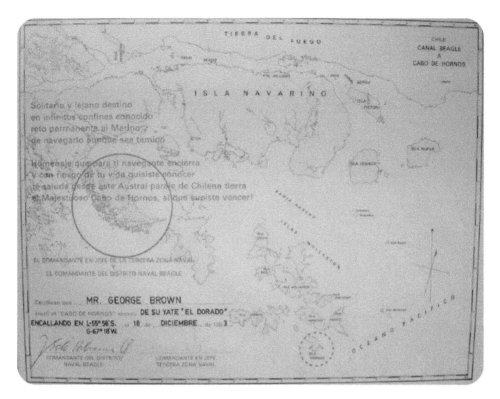

Certificate given to me by the Chilean Navy. They played a large part in my survival

Painting by Ian Hansen of Dorado approaching Cape Horn

CONSEQUENCE

Chapter 5
Back in Sydney

Back in Sydney after being away for two years required some readjustment. The first thing I discovered was that Alf and Clare had sold the marina and left Balmain to live at Alf's mother's house at Mascot. When I fronted at the Marina hoping for a berth, the new owner, Jeremy Lawrence, didn't like the look of me—I still had a two-year growth of beard, and hair down my back. He couldn't help me out, he told me. Next door, Cameron's was still Cameron's, but there was a new manager there, and he couldn't help me either. None of the people who'd been there when I left were still there.

What to do? I remembered that, down at East Balmain, next to the Sydney Ferries wharves and workshops, was Conway's boatshed. It wasn't a marina as such, but there was a slipway and some moorings. (I'd been on the slipway there once before when I couldn't get a booking at Alf's.) It was actually a better slip than Alf's; the track ran right down into deeper water, which meant that 'Dorado' could be slipped at almost any time, i.e., we didn't need to wait for high tide. I'd met old Ted Conway once before, with Frank, so I took a chance and went to see him. It was third time lucky. Ted rented me one of his moorings at a very reasonable price. I had to use the dinghy to get ashore, but that was no problem. We were back. Back at Balmain.

Meanwhile, back to Alf's old Balmain Marina, Jeremy Lawrence was a businessman, and the marina was just a sideline. He was a member of the Cruising Yacht Club and used to sail regularly, although I don't believe he ever owned a yacht. He was also a prolific taxi user, and I had picked him up several times over the years. He had found out who I was and what I'd done, and he tried to entice me back to his marina. I wasn't having it, though. His reputation in the taxi industry framed him as a complete arsehole, and I wasn't going to leave 'Dorado' anywhere near his grasp. He tried his bullshit on me in the cab, but I just laughed at him.

My taxi license had expired, but I had no trouble getting it back. They were always looking for drivers. Lots of people tried it, but very few survived. An experienced driver like me went straight in. It was tough for the first few weeks. The hurly-burly of city life and the cut-throat cab-game took a bit of getting used to, and the heat got to me, too. I'd been on a yacht on the water, with a breeze to keep me cool for the last two and a half years, and the heat of the land took me by surprise.

To circumnavigate the Southern Ocean non-stop in 'Dorado' would take about four months, more or less, I reckoned, and the southern summer was the time to do it. The days are long, and not quite as cold. The weather and sea conditions are probably slightly better. If I could leave Sydney during November, and get back before the end of March, that would be the most suitable time.

I had arrived back in Sydney in September 1980, having completed the circumnavigation. I was stoney broke and had to get back to work as soon as possible. It soon dawned on me that there was no way that I was going to be ready for the Southern Ocean that year; there was too much to do, and not enough time or money to do it. It wasn't a hard decision to put it off for a year, it was just common sense.

Dave Benson came to see me one day, down at Conway's. We'd met three or four times along the way. He had his own steel

Chapter 5 - Back in Sydney

yacht, which he'd built himself. We'd talked and discussed plans for the future. He knew I was planning to make some alterations to 'Dorado', and he was hoping to do the job for me. 'How did you find me down here?' I asked him.

He said he knew I'd be around Balmain somewhere. The rest was easy. So, we went to see Ted and booked the slip for a few days, and Dave did the job for me. 'Dorado' had a fin keel, and I wanted to extend it to a fuller keel in the hope that it might help with the broaching problem. It did help a bit, but the real problem was in the design and there was nothing I could do about that.

Dave had sold his yacht and was moving up the country to the Macleay River near Southwest Rocks where he had bought some oyster leases. I never saw him again after he left Sydney.

Conway's was a very quiet place, and that suited me fine, I was working six days a week and Sunday was a day of rest. That meant that I wasn't doing a lot of sailing, just an occasional Sunday and no major voyages. I kind of missed the social life of the two marinas, but of course everything had changed. Frank had taken his boat up the river to Mortlake and I hardly ever saw him. Hilda had sold her motor sailor and had left the area. Work was the most important thing at the moment.

In the cab game, you don't have many or any friends. It is a competitive dog-eatdog, cut-throat industry, and these nice guys that you meet socially will rob you in an instant out on the road. But if I had one friend in the cab game, it was Joe Grech. I met him early on, as one of a group of guys who used to go out together after we finished work at 3am. I'm not sure what the attraction was, except that I think some of the others used to bully him a little, although not when I was there. He was a very smart cabdriver and made lots of money. He was my mentor for everything taxi, and everything money, and we became close friends over a period of years.

I'd been back for a couple of weeks, and I was driving again, and I ran into Joe.

'George, it's you,' he says, shaking my hand. 'How are you? What's happening?

'I heard you were back,' he said.

We had a chat for a while about this and that, then he said to me, 'George, since you've been away there's this new game started in NSW.

'It's called Lotto,' Joe said, and went on to explain it to me.

'It's legal, and it's based on the American numbers game. You put your entry in, and the numbers are drawn on Monday night.

'I'm running a syndicate,' he said. 'Do you want to be in it?'

How could I resist, I became the fifth member of the syndicate.

So, the weeks and the months went by, and we couldn't win, not even a minor prize. One of the fellows got fed up with it and dropped out, and a couple of weeks later a second one went. That left the three of us, Joe, me, and Frank Pellegrini, who ran the service station where we started and finished our shifts. Pellegrini was a mad gambler, and unreliable, as they are. We were taking turns to buy the ticket (same numbers every week) and one week it was Frank's turn to buy. But hadn't shown the ticket to us, so I fronted him.

'Frank, have you got the ticket?' I asked him.

'Nah,' he said. 'I've left it at home.'

He didn't produce it the next day, or the next. Joe and I had

Chapter 5 - Back in Sydney

a get together to discuss what to do. We both reckoned he had not bought the ticket. So, what do we do now? I said, 'Look, I'm going to buy the ticket for myself, just this once, then no more. I'm finished.

'Do want to be in it?'

'Yes,' said Joe. 'Just the two of us, but Frank is out, OK.'

Monday came and I was working as usual, 3pm-3am.

The draw is about 8.30pm on Monday night, and the numbers were called as they came through on a radio station. I was sitting on Chatswood taxi rank; it was

8.30pm, the door opened, and a gentleman got in. The announcer was reading the numbers; the passenger was talking to me at the same time. I knew we'd won a prize, but I'd missed two of the numbers. As soon as I dropped the passenger off, I went to the nearest phone and rang the number that you can call to find out each week's results. I listened, and I listened again. I was having trouble grasping what I was hearing. We'd got every number, plus the supplementary! We'd won a major prize!

The feeling of euphoria was unbelievable. I felt like I was floating on air. I kept working for the rest of the shift, but I didn't do very much. Monday night is a quiet night anyway. At knock-off time, I grabbed Joe before the regular card game started.

'You got a moment?' I said to him.

'Yeah,' he said. 'What's happened? Did we win a prize?'

I said, 'Yeah, we won a prize…'

He was looking at me quizzically.

'How much?'

I said, 'First prize'.

'You're pulling my leg,' he said. 'Tell me again.'

'I tella you true,' I said.

It was starting to register with him, and he was starting to get excited. I said, 'Settle down mate. We got things to talk about. We gotta keep it quiet for the moment, and we gotta think about Pellegrini.'

'We think he didn't buy the ticket,' I said, 'but maybe he did. If he did, then we are entitled to our share of it. Let's just be sure.

'Anyway, it's still early days. I'll know more tomorrow.

'I'll give you a ring tomorrow before you go to work, OK?'

Next morning, I was woken by a commotion in the boatshed. It was the officials from the Lottery's Office, they'd been trying to find me, and the lady in the newsagency where I bought the ticket had pointed them in the right direction. So, the win was formalised there and then. I discovered that there had actually been six winning tickets, and by the time that all the sums were done, Joe and I were getting $85,000 each. (As a reference, an average priced house in Sydney at that time—1981—was $40-45,000).

When the deal was all done, the bloke said to me, 'Any questions?'

I said, 'Yes, can I ask you if any of the other main prizes was won by someone called Pellegrini with the syndicate name 'Desperadoes'?'

'I can't divulge names,' he said. 'Sorry,' but he was shaking his head.

'OK,' I said, 'that's good enough for me... Thanks very much.'

He gave me a phone number to ring, if there were problems, and they all left. I had to go up the street to ring Joe and bring him up to date. He was quite happy, and he agreed: keep it quiet for now, and don't say anything to Pellegrini; just keep an eye on him, and we'd know soon whether he'd won anything.

I soon worked it out that the money wasn't enough for me

Chapter 5 - Back in Sydney

to retire and never work again, but if I could keep it in one piece, and invest it wisely, it would look after me in later life. My first investment was with the bank, a five-year term deposit paying 13 per cent pa. (Interest rates were extremely high in that era. They were to become even higher later on.) Then there was share trading, and superannuation. Things were looking good in 2007, but then along came the GFC in 2008, which was a big hit, with a very slow recovery. Luckily, I'd kept working, and I was enjoying it. I was getting to be a bit of a junkie, but then along came covid, and the cases started to increase rapidly. Everybody was running scared, and business was no good. The sensible thing to do was give it a rest for a while. Rest became retirement a few months ago. The lotto win has looked after us. We're not millionaires, but we're comfortable, we live in a nice house in a nice suburb, health is not too bad, there are no major issues. Sydney is a nice place to live, with a nice climate, there is every facility that you could wish for. Australia has been good to me, and good for me.

I'm getting a bit ahead of myself here… We need to go back to 1981. I'd been almost ready to leave on my trip across the Southern Ocean when the lotto win occurred. That disrupted everything for a couple of months. Then I received news from Scotland that my mother had had her second stroke and had been taken into care. Also, my brother's marriage had broken up and he was divorcing. I put the sailing trip off for a year and got on a jumbo jet and flew to Scotland to be with my family. British Airways had just been privatised and were offering London return tickets for $1000. I stayed for Christmas and New Year and got back to Sydney in February.

With more money, I could plan a bigger and better trip, not

just a dash round the Southern Ocean. I wouldn't clear customs in Sydney; rather, I'd sail down the NSW coast, stopping maybe at Eden just north of the Victorian border, then on to Tasmania, stopping at a few places there before getting into Hobart after the boats had left after the Sydney-Hobart yacht race. From Hobart we'd cross to Bluff in the South Island of New Zealand (giving me a little taste of the Southern Ocean). Then I'd spend the best part of a year in NZ before taking on the Southern Ocean and Cape Horn. After Cape Horn we'd head up the Atlantic to the UK, and Scotland in particular.

It was important to keep working, to have a steady income, and not touch the invested money, other than in an emergency. 'Dorado' was rerigged, and we had some new sails made, a new mainsail in particular, with slab reefing this time; and a second yankee, a smaller one this time, especially for the Southern Ocean. My trusty Citizen watch would still be my chronometer, and I still had the ship's clock as a standby. It would be permanently set on GMT for navigational purposes and had a reliable second hand. I still had the same radio receiver, which had never let me down anywhere, getting the important time signals from WWV.

I hadn't done much sailing in a long time, just an occasional Sunday, and when an opportunity came along to do a weekend trip up to Port Stephens, I welcomed it. I'd been driving for a private owner for some time, Joe Manton. He was an ex-Navy man and was interested in what I was doing. We'd talked about sailing trips, and I asked him one day whether he would like to do a weekend trip up to Port Stephens. He thought it would be a good opportunity to celebrate her birthday with his wife, Colleen.

'It's Colleen's 40th birthday shortly,' Joe said. 'A trip like that would be something special.'

Yeah, I'm thinking this is August, and every Sydneysider

Chapter 5 - Back in Sydney

knows that, in August, we get the 'August winds', the boisterous westerlies that blow dogs off their chains and oysters off rocks. Anyway, we set a date, but I had to warn them about the 'winds.'

The day came, and the weather was OK; the winds were quite moderate, and the forecast was good, so off we went. We had a glorious reach up to Port Stephens, covering the journey in less than 20 hours. We spent a couple of enjoyable days up there, then it was time to return. The weather was forecast to stay fine for another couple of days, but there was a strong cold front coming through the Great Australian Bight, and we needed to be back in Sydney before it arrived.

We left Port Stephens around lunchtime the next day with a light nor'-east breeze and hoisted the spinnaker as soon as we cleared the heads. With the help of the current, we raced away to the south. The nor'-east breeze faded out in the late afternoon but was replaced by light nor'-west breeze of 8-10 knots. We kept the spinnaker up, but jibed, which meant we kept the same course, but we were now on a broad reach, and we held that until the early hours of the morning when we were off Barrenjoey (Palm Beach). At that point, the breeze shifted a little more to the west and increased to 10-12 knots. That meant that the spinnaker had to come down and be replaced by the genoa. We continued to make great progress and came around North Head just as the sun was rising. It was windward work from there and the sheets were 'hard on'. We held the starboard tack all the way up the harbour to Point Piper, then switched onto the port tack and held that all the way to Milsons Point. That was enough. That winter morning westerly was perishing cold, and the thought of tacking back and forth under the Harbour Bridge with all the ferries coming and going from Circular Quay was too much. We dropped the sails and motored back to Balmain.

DESTINATION CAPE **HORN**

It had been a very successful weekend, enjoyed by all, and especially me. I really needed that weekend; it just set me up. I was near ready to leave on my second circumnavigation.

Chapter 6
Cape Horn

I got stuck into the preparations with renewed vigour, and by early November we were ready to go. There was a nor'easter forecast, and we picked it up as we passed through the heads (early November 1982). The northerlies held for two days, but became a bit unsettled in the end, and there was a thunderstorm. I was really trying to get to Eden, but the morning after the thunderstorm a sou'-west change came through, the breeze started to build, and the going was starting to get tough. We were just off Bermagui at the time, so we took the easy option and shot in there.

It was the right option. Bermagui is a charming little fishing village, and we were safe and secure while the Bass Strait gale blew itself out. A few days, later the wind swung northerly again, and off we went, by-passing Eden, and out into Bass Strait. I was hoping to get across the strait and well down the east coast of Tasmania before the next change came through. Bass Straight was on its best behaviour, and we had moderate winds all the way across and down the Tasmanian coast.

We were approaching Schouten Island late in the afternoon when the breeze started to freshen. I'd been hoping to get to Triabunna before dark, as there was a tricky access to the creek through mudflats, but that wasn't going to be possible, so I anchored off the beach near Orford. It was quite safe as the breeze was coming off the land. I could see a hotel right on the beach, and through the binoculars I could see a cricket match on the TV. I launched the dinghy and rowed ashore, got some take-away food, and watched the cricket. The next day it was an easy move up to Triabunna, arriving at high tide. It was crowded in the creek, and I was lucky to find a visitor's berth for a few days. Triabunna is a timber town, and any visitor is suspected of being a 'greenie'. I was quite happy to leave there when the time came.

Our next move was to get around Tasman Island, and up into Port Arthur, a day sail. (Port Arthur was, of course, a penal colony in the early years. The people sent there were the worst of the worst criminals. They built the gaol, and the prisoners lived there, but the place was a misery for everybody who lived in that area.) I was about to drop anchor when a guy on a nearby mooring called out to me.

'There's a mooring here,' he called. 'The owner's away… He won't be back till February.'

'Oh, thanks very much,' I said. 'I'm not going to be here that long,' and picked up the mooring.

It was only a couple of days later that I was on deck, and I saw a large launch coming round the headland. Oh, oh! I thought. Here comes trouble. And sure enough, he came right up alongside, and demanded his mooring back. I obliged straight away and went off to anchor nearby. The holding was good; it was a perfectly good anchorage. It was what I should have done in the first place.

My plan had been to stay in Port Arthur until early January, when all the Sydney to Hobart boats had cleared out of Constitution Dock. I wanted to get up there myself to prepare for the next leg of my voyage. Port Arthur had a depressing pall hanging over it. It had been an absolute hell hole for many years, for everybody, not just the prisoners. Nowadays, there are busloads of tourists visiting there for an hour or two, then they move on. I wouldn't like to live there permanently. (In 1996, a local man, Martin Bryant, went there with his gun, and pocketsful of ammunition, and started shooting visitors. He killed 35 and

Chapter 6 - Cape Horn

wounded many others before he ran out of ammo. Now he lives in solitary confinement in a mental institution and will stay there for the rest of his life. A sad footnote to the Port Arthur story.)

Looking at my chart, I could see another likely anchorage not far from where we were, a small bay which took a dogleg to the right. It looked very sheltered, if a little shallow at low tide. There were apple orchards there, and a small settlement, Nubeena. It looked like a good option. It took only a few hours to get around there and anchored safely. I had the place all to myself. I launched the dinghy and went ashore and found a shop. It had all the fresh stuff that I needed; any major shopping could wait until we got to Hobart. I had nearly a week in Nubeena studying all the charts of my next destination, New Zealand. I was planning to spend the best part of a year in NZ before taking on the Southern Ocean, and Cape Horn, and all charts had to be studied in great detail. Study them carefully now, then, when they were required, I was already familiar with them.

I moved up to Hobart and Constitution Dock around January 10 and, as I'd hoped, it was deserted. It was a very safe and secure place, and I had a very pleasant stay for a couple of weeks. The days went by and suddenly it was the end of January, and I had my clearance for Bluff in the south of New Zealand's South Island. The Southern Ocean was on its best behaviour; it was an uneventful trip, and we ended up motoring down Foveaux Strait in an almost flat calm.

Bluff harbour is quite industrial, but there was a fishing fleet over in one corner. I noticed a fisherman beckoning me over. I rafted up alongside him.

'Our fishing season starts in two weeks,' he told me.

'I'll be gone long before then,' I said.

'OK then, that's fine,' he said. 'You can stay here for now.'

I thanked him.

There wasn't a lot to do in Bluff. I took a ferry ride over to Stewart Island just for something to do. I visited the pub a couple of times; very basic, it would be easy to get into trouble there. The Southern Ocean had come to life again, and the westerly wind was howling down Foveaux Strait. I took a bus trip up to Invercargill. It was cold, the wind was blowing, and the streets were deserted… nothing to see here. It was time for 'Dorado' and me to get going again, up north where the weather is warmer.

Back in Bluff, I had a chat with the fisherman about the weather.

'That wind will blow for weeks,' he said. 'If you want to go, go now.'

Next morning, I got the engine going (essential for manoeuvring in restricted areas), and the sails ready to hoist. A small headsail, and the mainsail with reefs already tied in. Let go the mooring lines, and away we go. Almost as soon as we let the mooring lines go, I noticed that the engine was overheating. Panic stations! I had to turn the engine off, and get the sails up, and sail out of the harbour in gale force conditions. The next hour was very busy; getting the boat safely out of the harbour, and the off-lying dangers, trimming the sails, and setting up the Aries selfsteering gear to keep us on the correct course.

The wind was sou'-west force 7-plus. We were cruising along at 6-7 knots. Things had settled down a bit. Time to have a look at the engine. I knew what the problem would be; it had happened before. The cooling water inlet was a very narrow pipe, and it had no filter. Just occasionally, a foreign body gets in and jams the pump valve. So, we close the inlet valve, dismantle the

Chapter 6 - Cape Horn

water pump, clean out the foreign body, probably replace the little plastic valve, and reassemble. I'd done it several times before, but it always seems to happen at an awkward time.

I had mulled Dunedin as my next stop. It was several miles up the river, and I didn't have any compelling reason to go there, but right at the heads where the river enters the ocean there is a small bay, where Robin Knox-Johnston had anchored momentarily on his epic voyage round the world. I thought maybe I could anchor there for a night or two, just to rest up. Navigation was just by dead reckoning, there was thick cloud covering the sun, but there are navigational lighthouses along the coast, and one in particular on the headland at the entrance to Dunedin's estuary, Port Chalmers. We were off Port Chalmers in the middle of the night, and conditions were still quite rough. I decided it was safer to carry on than attempt to enter some place where I had never visited, in these conditions. Oamaru was the next port up the coast. Not all that far, we'd be there in daylight hours. It was a better option. The weather and sea conditions eased a bit over the next few hours, and the sun even appeared.

The approach to Oamaru was like a Sunday sail. We came around the breakwater and entered the harbour. It was quite crowded, obviously a fishing port. There were lots of fishing boats on moorings. There was a fisherman's co-op, which looked very busy. I decided to anchor near the moorings, go ashore and find the harbourmaster. The message was: This is a fishing port, and very busy. It's not that you're unwelcome, but we don't want you to be getting in the way.

'We'll let you stay for a week for free,' they told me, but 'if you stay longer, we'll have to charge you.'

That was all fine by me. I was only going to stay a few days. The boat was fine where it was, and there was a small beach in the nor-west corner of the harbour, where I could land my dinghy. Leaving your dinghy unattended on a public beach is not recommended, but sometimes you have no choice. There were two or three occasions on the circumnavigation where I have left the dinghy unattended and come back to find the local kids making free with it. I think I went ashore three or four times in Oamaru, and there was no problem. We were there for three days, and that was enough.

I'd been thinking of moving on to Timaru next, but I was told, no, don't go there; it's smaller and even more crowded than here.

'Go to Akaroa,' they told me. 'That's a better place for you. You'll enjoy Akaroa.'

I'd noticed Akaroa, when planning the voyage, but I didn't take it too seriously. But I'm glad I went there; the place is an absolute gem. Akaroa is on the south side of the Banks Peninsula. A long fiord runs inland about 15kms. There's high ground on both sides until near the end, where it opens out into a beautiful bay. Absolutely safe and secure. It had been a popular place for whaling ships, back in the day, especially the French. At one time the French tried to claim all of the South Island, but they were too late, and the British beat them to it. Today, Akaroa remembers its French heritage and celebrates it. I stayed in Akaroa for 10 days. The weather remained fine and warm, and I enjoyed that. I felt a little sad to leave such a beautiful place. Maybe we'd come back again, sometime.

The next leg took us around Cape Banks, and into Lyttleton, the harbour for Christchurch. It was smaller than I expected, and quite crowded. The yacht club was obvious, and as I looked over that way, I could see a fellow waving to me. I went over, hoping he might have a visitors' berth. Some hope! As soon as we got

Chapter 6 - Cape Horn

close, I could see his face fall.

'I'm terribly sorry,' he said. 'I was expecting a yacht this morning, but it wasn't you.'

Then he said, 'No, I'm sorry we don't have any visitors' berths, but that old wharf over there,' he said, pointing over to the other side of the harbour 'has nothing on it at the moment. You'll be safe enough there. ' The wharf was timber, built on timber piles, and had been built for bigger vessels than 'Dorado'. Even at high tide, it was difficult to get up onto the wharf from the deck, and at low tide it was well-nigh impossible. Coming back on board wasn't so bad. From the wharf, I could reach out and grab the shrouds and shimmy down onto the deck. It was far from perfect, but I wasn't staying long. I was determined that I was going to get into Christchurch and have a look around, if I did nothing else. The weather stayed fine, and I did get the bus into town. It was like a big country town, the population at that time must have been around 500,000. It has the reputation of being a very churchy town, and I saw plenty of them. But there doesn't seem to be a lot of industry. It's the administrative centre for the NZ district of Canterbury, and it has a famous rugby team, the Crusaders. That was Christchurch; most of what I saw was ruined in the earthquake of 2011, but life goes on, doesn't it.

Back on board, I decided to make the most of the fine weather and get going. The next move could have been to Wellington, but I had my doubts, and I didn't want to be disappointed, as I'd been in Lyttleton. I'd been reading about the Marlborough sounds, and Picton in particular. That sounded attractive, and it's what I decided to do. The distance from Lyttleton to Picton was a bit over 150nm, which meant a two-day trip, and of course it meant going almost all the way up Cook Straight, that notorious stretch of water, where gales are very common. In the end it turned out to be a light weather trip, and we ended up motoring up Cook Straight.

The only drama occurred after we entered the sounds. Going along, minding our own business, I noticed what looked like a red stain on the water ahead. As we got close, I could see that they were tiny red fish shoaling close together. They were krill. As we started to move through them, I suddenly realised what was about to happen: they were going to block the water pump, weren't they! I made sure that I got right through the shoal, and well clear, then stopped the engine, and cleaned out the water pump. It was jampacked full of krill. I saw two more shoals on the way down to Picton, and I went around them this time.

Picton was a great place to stop. I enjoyed the hell out of it. I also discovered that there was a regular ferry service from Picton to Wellington. If I caught the early service from Picton, then the late ferry back from Wellington, that would give me about five hours in Wellington… plenty of time to scratch around and see whether it was worthwhile going there with 'Dorado'.

The day that I did the trip was a fine weather day, but when we got out into Cook Strait, there was a nor'-west wind blowing at 30 knots. It would have been a tough day for me and 'Dorado' trying to reach the sounds in that wind. As soon as the ferry berthed, I walked around to the yacht club. It wasn't very far. The fellow in charge of the marina was sympathetic, but he couldn't offer me anything right now.

'Give me a ring in a couple of weeks,' he said. 'You might have a better chance.'

'What about the rest of the harbour,' I asked him. 'Is there

Chapter 6 - Cape Horn

any safe anchorage?'

'In a word, no,' he said. 'It's not recommended. The ground here is not good holding, especially considering the weather we get... Those boats that you can see over to the east are on specially reinforced moorings... Perhaps if you knew someone over there, you might get to use one of their moorings.'

What I was hearing didn't surprise me. It just confirmed what I suspected: that Wellington was not the place for 'Dorado' and me.

There was another option that I could consider as long as I was in the sounds. The town of Nelson was said to be part of the Marlborough sounds district, but it was right round on the Tasman Sea coast. By road, the bus would take about an hour, but by sea it would take 2-3 days in possibly iffy weather. So, I made a day trip of it on the bus to check it out. The town itself looked very nice, but the facilities for visiting yachts were very limited. The boat harbour was quite small. When I was there, it was low tide, and the harbour had partially dried out. I'd seen enough of Nelson. I wasn't going to go there. It was time to move on, to get up north before the winter set in.

As soon as I got back to Picton, I started studying my charts and planning our voyage to Napier. The distance was about 250nm, possibly a three-day trip. Get out into Cook Strait, around Cape Palliser, right at the southern tip of the North Island, then up the east coast of the North Island until we get to Hawkes Bay, and Napier, at its southern end. I had a great time in Napier. The yacht club was friendly and accommodating, and there was a lot of socialising. At one stage, someone in the yacht club introduced me to a fellow from Wellington, who'd come up to Napier to buy a boat. He was looking for someone to crew with him to get the boat back to Wellington. He was even offering money. Nobody would go with him, though, and I knew why they wouldn't go. The bloke was gay. I'd been a Sydney taxi driver for too many years not to know. OK, I thought, I'll take your money, and get to sail into Wellington Harbour. So, we talked it over, and agreed on money, etc.

The boat was a Hartley design, well known in NZ, of ferro-cement construction, more of a motor-sailor than an outright sailing boat. The gear all looked OK. We sailed all the way down to Cape Palliser. As we approached the Cape, the breeze started to increase, and was going to increase as we entered Cook Strait. We put a reef in the mainsail and started up the engine. This was going to be tough going. The wind was from the north, up to 30 knots. We held the starboard tack for a long time then tacked back to the north. We were half-way to Wellington. So starboard tack again for three more hours, then port tack back towards Wellington… not quite there… starboard for an hour, then port, right to the entrance to the channel, drop the headsail, and motor through the channel, and into the harbour. The wind eases a bit, and the worst is over.

He'd been pretty quiet all the way up the Strait, but at least he didn't get seasick. He had a mooring over on the eastern side of the harbour. He'd been making phone calls, and somebody was there to meet us with a dinghy. It was late in the afternoon, and I wasn't going back to Napier that night. A good shower, a meal, and off to bed. In the morning, a bite of breakfast, then into the city to catch a bus back to Napier.

Back in Napier, it was time to start planning my next move. I wanted to get to Tauranga, and maybe stop at Gisborne on the way, 85nm north. The locals at Napier didn't think much of

Chapter 6 - Cape Horn

Gisborne, but I was going to stop there anyway. The Napier locals were right. The harbour was very small, facilities were very limited, and there was no yacht club. Officials didn't want me to stay any longer than necessary. I ended up staying for three nights. The only worthwhile thing that happened there was when I took a walk in a local park and saw my first live kiwi!

The voyage from Gisborne to Tauranga, almost 200nm, should have been fairly straightforward: north up to East Cape, then west all the way to Tauranga. When I reached East Cape, however, a gale warning was issued. The people at Napier had warned me about East Cape, and they'd told me about Hicks Bay as a good place to shelter, so I headed in there and joined about a dozen fishing boats, and together we rode out the gale. About 24 hours later the gale had just about blown itself out, and I got under way again.

It was about 100nm to Tauranga, and the course was almost due west.

Keep clear of White Island, the fishermen yelled to me, as I sailed away. White Island was an active volcano, but a volcano filled with water. This meant that, when it vents, what comes out is steam, and poisonous, super-hot fumes. In later years, White Island has become a bit of a tourist attraction, despite the danger. In 2019, there a large group of tourists were on the Island when it vented unexpectedly. If I remember correctly, 27 were killed.

It was fast sailing to start with, but the weather steadily moderated, and almost faded out and by the time we got to Tauranga, we were motoring. We had passed White Island in the middle of the night. There was steam coming out of the volcano, but no red glow. The entrance to Tauranga is actually a river mouth, dredged to accommodate large vessels. It's one of NZ's most important trading ports. The entrance to the estuary is narrow with strong tidal currents, and it is best to enter with a rising tide. We were lucky; we arrived at the right moment and shot up the river at high speed. My sailing directions told me to follow the river right up to nearly the railway bridge, and there on the starboard, would be the yacht club.

I could see the railway bridge quite clearly, and that building with the floating pontoon would be the yacht club. So, we tied up to the pontoon, expecting someone to come and to see me. No one came, so I went to find somebody. I walked into the building, which I thought was the yacht club. There was nobody there, but I could hear voices coming from somewhere. I walked a bit farther, and there they were in a room, about 10 of them, and there seemed to be some sort of party going on. They were all looking at me, so I said, 'I'm sorry to butt in, but I thought this was the yacht club.'

They all started laughing.

'No, this is not the yacht club,' one of them said. 'They moved from here five years ago. This is the J Walter Thompson advertising agency. We've taken over the building.'

'Oh, I'm terribly sorry,' I said. 'I didn't know. I've tied my yacht up to the pontoon.'

'That's OK,' the fellow said. 'No problems. You're welcome here. You can stay as long as you like.

'Have a beer!'

What a stroke of luck that was. Tauranga was just about the nicest place I visited in NZ. They didn't charge me for my berth on the pontoon, and they set me up with a job, and arranged transport to and fro.

The hinterland of Tauranga to the sou'-west is the main Kiwi fruit producing area in NZ. The fruit stays on the tree until it gets that first little nip of frost in the autumn, then it's all hands-

Chapter 6 - Cape Horn

on deck to get the fruit in and stored before it's damaged. I'd arrived in Tauranga in early May, just at the time that the call went out for workers. I'd be picked up on the street by a car with three people already in it, and we drove for half an hour out to a large farm.

I was given the job of driving a tractor, towing a trailer, delivering picked fruit from the field up to a large storage shed. The trailer had a swinging axle. Boy, did I have some fun with that thing. They were taking bets that I was going to stuff up and lose a load. Every year, they said, every year somebody stuffs up and loses a load. Well, they were disappointed, because I never lost a load. I came close a few times, and they all stood around and cheered, but I never lost a load. The money wasn't great, but that wasn't all that important, I really enjoyed getting out and about, and meeting new people. It was good fun.

When it was finished, and we were all about to leave, a fellow approached me and introduced himself. He was a contractor, he said, and could guarantee me work for the next six months, better money than here. Would I come and work for him? I was taken aback. It was a great offer, but I was planning to get moving up the coast very shortly. Sadly, I had to refuse the offer, Dorado, and Cape Horn came first.

Just a couple of short stories, if I may…

A couple of days after I arrived in Tauranga, I walked up the street to get a soft drink or something. There was a small grocery store on the corner, and I went in. There was an elderly man with white hair, behind the counter. He was kind of staring at me, like he was ready to pounce. I walked up to him and said, 'Good afternoon, sir. I'd like to buy a bottle of soft drink, thanks very much.'

He was still staring at me as he said, 'Where do you come

from?'

I heard the Scottish accent loud and clear. It was a border accent. You couldn't mistake it. I said, 'I was born in Kelso.'

'I was born in Selkirk,' he said.

He was a good deal older than me, so we didn't have all that much in common, but we chatted for a while. He let it slip that the people at the advertising agency had told him about me, and he must have been expecting me to come into his shop.

'Don't forget to do your shopping here,' he told me, as I left the shop.

I was doing my washing one day on board 'Dorado'. It was cold, and the sou'-west wind was whipping across the water. I was using a bucket on deck. It was a good drying day, and a good opportunity when I wasn't doing anything else. I was getting stuck into this miserable task when I noticed an old gentleman come onto the pontoon and start walking towards us. I kept my head down, thinking, no, not today mate.

I'm busy today, can't you see… come some other time, and we'll talk. But he was determined, and he wasn't going to be denied.

'You come from Sydney,' he said.

'Yes,' I said. 'I come from Balmain.'

He looked at me for a moment, then he said, 'If you come from Balmain, you must know Alf Taylor.'

That stopped me. Alf and Clare Taylor were the owners of the marina where Dorado was berthed, and where I had lived on board for all these years. They'd been very good to me in many ways for all the years that I'd been there.

'Yes,' I said. 'I know Alf Taylor.'

'Has he still got the sailing school?' was the next question.

Aha, the story was coming together now. I knew who this

Chapter 6 - Cape Horn

guy was. Alf had spoken about him a number of times. Alf and Clare did have a sailing school back in the day, where merchant seamen could go and study for the various tickets necessary for promotion in the merchant marine, right up to Master. Alf remembered this guy as the last person to study and achieve the master's certificate for sailing ships. Alf referred to him as The Kraut. He did have German heritage; I could tell by his accent.

His parents had lived in Tonga before the first world war, when Tonga had particularly strong relationship with Germany, and he had spent most of his life around the Pacific.

'Are you going back to Sydney?' he asked.

'Eventually,' I said.

'Please pass on my regards to Alf and Clare,' he said.

He came again on another day, and we talked about sailing ships. He had crewed aboard the clipper ships, which delivered grain from South Australia to Europe, right up until WW2. He had gone round Cape Horn on numerous occasions and experienced the worst of conditions.

'Are you planning to go around Cape Horn?' he asked me.

'It's the long way back to Sydney, isn't it,' I said.

He wished me good luck as he left. I think he knew perfectly well what I was planning to do.

My next move was up to Auckland. I was going to stay close to the coast, and day sail it. My friends in Tauranga had given me a list of anchorages safe from the westerly wind, and I was going to hop from anchorage to anchorage. We were into winter now, and it was cold, and the days were short. There was no rush, however, and although the westerly wind was very strong at times, the weather stayed fine, and there was no rain or fog to make navigation difficult.

The day that we came around the top of the Coromandel peninsula the westerly was blowing about 25 knots across the Hauraki Gulf, and we were close-hauled beating into it tack on tack. It was late in the afternoon, and it was looking like we were not going to get into Auckland before dark, when I noticed this beautiful white beach on Waiheke Island. Just the thing, I thought. We'll drop anchor off the beach, and we'll be out of the wind, and snug for the night. In the morning I launched the dinghy, and went ashore, found a shop, where I got some fresh bread rolls, eggs, and milk. The wind was still pretty strong, so it was an easy decision to stay for another day.

I didn't want to stay in Auckland for very long, just another grubby city, and expensive, too, probably. There was a big new marina just up the river a bit, but I didn't want to go there, unless I had to. As I approached the city, I could see that most of the finger wharves were vacant. I picked one and went right in there close to the roadway. The wharf was quite high like the wharfs in Lyttleton, but this time there was a ladder. I

berthed next to the ladder, and prepared to go ashore, and see the harbour master.

As I reached the roadway, there was a guy walking towards me.

'G'day, he said. 'Is that your boat there?'

'Yes,' I said. 'I was just going to see the Harbour Master.'

'I'm from his office,' the fellow says.

'How long do want to stay?'

'Just a few days,' I said, 'then I'm off again.'

He said yachties like me were not normally allowed to stay on those wharves, 'but it's the middle of winter, and very quiet, so you can stay for a few days.

Chapter 6 - Cape Horn

'We have a ship due in next week, so you'll have to be gone by then,' he said.

'Thanks very much, mate,' I said. 'Will there be any charge?'

There would not be a charge, he said, 'but I'll give you a tip—this is an iffy area, so don't go away and leave your boat unattended for too long'.

'I got it,' I said. 'Thanks for the tip.'

I didn't do much in Auckland except send off some greeting cards, and a bit more planning. I didn't want to leave on the Cape Horn voyage before early November, so I still had some time on my hands. I wanted to get out of Auckland soon, and visit Great Barrier Island, then up to the Bay of Islands for a while, before coming back to Whangerei to prepare for Cape Horn.

Great Barrier Island was very peaceful after Auckland. The first anchorage was a bit rolly, caused by the persistent westerlies. The second attempt was better, but I had a long row in the dinghy to get to where the shops were. All the locals had an outboard motor to go with their dinghies. I could understand why. There were some good walks ashore, and I needed the exercise.

Great Barrier Island to The Bay of Islands should have been a day sail, about 90nm, but the westerlies let me down, and I struggled to get to Russell before dark. Kiwis talk about the Bay of Islands in the same way that Aussies talk about the Great Barrier Reef, and I was surprised to find that it was a comparatively small area. Nevertheless, it's a beautiful area with lots of secure anchorages. I spent a week there cruising around.

I was cruising around one day when I saw a timber wharf. It was unusual in that there were no boats tied up to it. Neither did there seem to be any boatshed or house associated with it. There was what looked like a freshwater tap at the end of the wharf. I wondered whether anyone would mind if I filled my water tanks up. There was nobody around, so I quickly put about and tied up at the wharf and got the water jerries out. It didn't take long to fill them up, and I was starting to put them back on board when I noticed somebody coming down the wharf. Oh, oh, I thought; here comes trouble. It was a lady, probably in her sixties, and she spoke with a beautiful English accent.

'I hope you're not thinking of berthing here,' she said. 'Our yacht, Wanderer IV, is normally here. It'll be back tomorrow.'

'I'm just going now,' I said. 'I just wanted some water. I do apologise for not asking.'

'That's quite alright,' she said. 'Good luck!'

As I motored away from the wharf, I realised I'd just met Susan Hiscock. Eric and Susan Hiscock were the doyens of the cruising world. They had written many books, of which I had at least three. My circumnavigation was based on their voyage in Wanderer III. I'd heard that they'd settled in NZ. I should have paid more attention.

I made my way back to Russell and anchored off the beach. There was a good shopping centre ashore, and I wanted to send some postcards, and other odds and ends.

Winter was over, and it was time to get down to Whangārei to prepare for Cape Horn. I just wanted to make one stop, at Tutu kaka. It had been recommended to me as a beautiful spot on a beautiful coast. When I arrived there, I found a marina in the place where I was expecting to anchor. OK, I thought, so I'm here now, let's just take a marina berth for a few days, and enjoy it for what it is. The place is probably heaving with people in the summer months, but this was the end of winter and I had it all to myself.

Chapter 6 - Cape Horn

I thought one day would be enough to get us to Whangārei, but I misjudged it. After we got around Busby Head, and Marsden Point, the river opened out into a large shallow bay. There was a marked channel, but of course that is not good enough when you are going tack on tack into a fresh westerly. The sensible thing was to anchor for the night, not too far off the channel in water that was two metres deep at low tide. Next day, we went right up the Hatea River into the middle of

Whangarei. We got a mooring, fore and aft piles, which was fine. I'd just have to use the dinghy to get ashore.

Whangarei was a good place to prepare for a long voyage. It was a customs port and had every facility we could wish for; lots of shops and bonded stores. There was a slipway about 200m down river from us, which I visited early on, and booked 'Dorado' in for three days, for two coats of antifouling.

There were lots of sailing people around, and so there was a bit of socialising. I was in Whangarei when Australia II won the America's Cup. What a day that was! I listened to it on one of the ABC shortwave frequencies and celebrated with the Kiwis. It would have been a great day to be driving a cab in Sydney.

The time passed quickly, and suddenly it was November. I didn't want to start my voyage on a Friday (it's bad luck to start a voyage on a Friday). On the Thursday, I was rushing around fixing up last minute details, and I was ready to go by lunchtime, except for one thing: the guy from the bonded store hadn't arrived with my duty-free liquor. I couldn't leave without that. Happy hour in the evening was a most important part of any day. He finally arrived about 4pm, just about enough time to get out to sea and clear of land before dark.

DESTINATION CAPE **HORN**

As we came around Busby Head, we encountered a moderate nor-east breeze… great, set the boat up, trim the sails, set up the self-steering gear. We're on a shy reach, course east-sou'-east, speed 5kts. Have a cup of coffee but stay on watch. Fishing boats can be a problem this close to the coast, not to mention other shipping. The plan is to follow the great circle course, which means an initial course of south, which changes slowly to the east, until you get to 56 degrees south, when your course will be due east (56S is the latitude of Cape Horn). It's the shortest way to go (don't be fooled by the Mercator projection used by the official charts).

Our initial course was east-sou'-east because we had to clear East Cape before heading south. We then had several days of light conditions. 'Dorado' goes best and sails fastest when she is reaching, and so in the light conditions I'd set her up to try and make the best of the conditions, but progress was slow. Whangarei is about 37.5S, and we still hadn't made it to 40S yet. I'm listening to the weather forecast… a Southern Ocean cold front was crossing NZ. This is what we wanted. Already the wind had swung to the nor east and was freshening. Over the next day, the wind went from nor'-east to north to nor'-west to west and freshening until the front came through. It got up to about 40 knots at the most, then backed off to sou'-west 25-30 knots. We got days of fast sailing out of it, and were well past 40S, and could hope for more strong westerlies, but it was not to be.

The wind staved south for a couple of days, easing all the while, and we did make some good progress in that time, the breeze eased right away, and the barometer was rising fast, and it was obvious that a big zone of high pressure was building behind us. The breeze backed to the sou'sou'-east and started to increase. That was consistent with the approaching high, and it

Chapter 6 - Cape Horn

was telling me that I wasn't far enough south yet. With the breeze at sou'-sou'-east, I could make a bit better than due east, and I decided to live with that for the time being, hoping that the breeze would shift.

The breeze did shift to the sou'-east, and it increased to about 30 knots, and we were pounding through short seas, with spray flying everywhere. I had hoped that it would have shifted to a bit north of east, but this was insane. I'd made a mistake. I've read numerous accounts by people who have travelled this route, and a surprising number of them get trapped in this same position. The answer is simple. Make your way south, until you get underneath the high, and that is where you will find the westerlies. It took me four days of bashing into this sou'-east gale; the best course I could get was sou'-sou'-west, and by the time the gale eased, and shifted to the nor-east we were down to 50 degrees south. What a relief, the wind had eased, and the seas had eased, and we were making fast progress on a course of east-sou'-east. Cape Horn is 56S, so east-sou'-east is a good course for now.

This little misunderstanding had probably cost us a week, so we had to get on with it. We were at 50S, and we didn't have the westerly winds yet, but we had a sou'-west swell, and it was building, so much so that in the trough between the waves we didn't have any wind, and 'Dorado' was left rolling around with sails flapping, potentially damaging the sails and other gear. Then as the next swell approached the wind came back and increased to about 40 knots as the crest passed under us. This was also putting severe pressure on all the gear. The worst of this lasted for about 24 hours, by which time I had just about gone potty. Finally, the wind shifted to the nor'-west, and increased a bit, and all the drama settled down.

DESTINATION CAPE **HORN**

Over a period of hours, the nor'-west wind freshened, and eventually got up to about 40 knots. A bit of a squall came through, and it shifted to west then sou'-west about 45 knots, then started to ease. We were in the Roaring 40s, and this type of weather would stay with us all the way to Cape Horn, about 3000nm away.

Navigation had been difficult for the last week or so. It had stayed mainly cloudy, and the opportunity for getting accurate sun sights was very limited. Trying to get accurate sights when you're bashing to windward in 30-plus knots of wind with spray flying everywhere is very difficult; even dead reckoning can be a bit iffy. But now that we were travelling more upright, and there was much less spray around, and much clearer skies, I got stuck in, and it wasn't long before we were up to date.

In 1983, GPS did not exist. There was a fairly new system called Satnav, which was suitable for commercial shipping, and relied on five satellites positioned above the equator. Some small vessels had them and were happy with them. I was advised that for a trip down the Southern Ocean they might not be very reliable, and I decided to stick with what I knew. Sextant and nautical almanac had seen me right round the world, and I was confident it would do me again.

Taking sextant sights on a small boat in a seaway is a skill, and practice is what hones that skill. Getting a horizon is difficult in the Southern Ocean where the massive swells follow one another, and the only chance is on top of the swell, which gives you a few seconds before you start descending the other side. A good chronometer is essential. Nowadays, most of the watches that you can buy in a jeweller's shop are fine, especially if you check them regularly with WWV. An error of one second means an error of on nautical mile in your navigation. My Citizen watch,

Chapter 6 - Cape Horn

bought in Fiji in 1975, had gone round the world with me, and was still keeping excellent time. The routine was: morning sun sight for longitude; local noon sun sight for latitude; afternoon sight again for longitude. I keep a running log, which I update every hour with course, distance, barometer, weather, position, and any other relevant information available at the time.

We'd been in the Roaring 40s for a week now, making great progress— 170-180nm per day. It was rough, but it was fast. I'd told the folks at home that I estimated that I'd be past Cape Horn by Christmas, and we looked to be on schedule for that.

I was checking the accuracy of the compass every day. The compass has a needle which runs through its centre. When the sun is shining, the needle throws a shadow across the compass. Take the reading of the shadow, then go back 180 degrees. That is the sun's bearing. Now, enter the nautical almanac, and look up the time to the nearest minute. That gives you the true bearing of the sun for that time. Compare the two figures allowing for magnetic variation, and you will see how accurate your compass is. Dorado is a steel boat, and the compass has to be set up very carefully and compensated for the boat's own magnetic field. It doesn't take much to throw this out. We were just entering a part of the world with very high magnetic variation, and the compass was going to be looking funny anyway.

The days were long and getting longer. By the time we reached Cape Horn, we would have daylight for 22-23 hours a day, so we should be able to eyeball any dangers that might threaten us. It was cold; the temperature gauge in the cabin said 6 degrees Celsius, and the ocean was 4C. I was wearing several layers of warm clothing, starting with woollen long-johns, with my wet weather gear over the top. I also had a pair of leather boots that I'd bought specially for this voyage, and soft leather sailing gloves.

I had plenty of food. I try to keep it simple. I have a small two-burner kerosene stove set up on gimbals. Porridge for breakfast; coffee out of a tube; tins of soup, very refreshing at any time of day; a variety of meat tins for the evening meal; heaps of oranges, potatoes, and onions. All of these have to be picked over regularly. Lots of fresh water—100 gallons. It's possible to collect fresh water at sea, but it doesn't rain a lot in the open ocean. You need relatively calm conditions, and you need to let it rain for 20 minutes or so to wash the salt off the sails, before you start collecting the water. And last but not least, a goodly supply of rum. Happy hour is an important part of the day. After the evening meal, it bucks the spirits up before the night engulfs us.

I haven't mentioned my companions yet. They deserve a mention (everybody else mentions theirs). There were two albatrosses. They'd joined me when I arrived in the Roaring 40s and stayed with me to Cape Horn. There was big daddy with his massive wingspan, continually gliding among the waves, and swooping back towards the boat, his eyes fixed on me. The other one was smaller, and younger, maybe his girlfriend, or even daughter. She didn't have the stamina of big daddy, and you'd often see her just sitting in the water preening herself. I never saw them eating anything, and they weren't interested in anything that I offered them. But they were company, and I enjoyed their presence. Who knows, maybe I was company for them, too.

One day, I had the radio on, checking my watch with WWV, and after the time check I was fiddling around with frequencies. Suddenly, an AM station came through, loud and clear. It was from a town in Texas in the US. It was morning in Texas and

Chapter 6 - Cape Horn

people were having breakfast and going to work. They were a raunchy bunch in this town, regaling each other with last night's conquests. I was fascinated with it and kept listening to it. It lasted for about 20 minutes or so before fading out. A fluke of nature, they brought something into my life, and I remember it well, 40 years later.

I'm a bit of a loner by nature; I always have been, and it doesn't bother me in the slightest to be on my own. When you're single handing, there is always plenty to do. Just because there is a self-steering gear doesn't mean you can set and forget. It needs monitoring and adjustment from time to time; sails need to be trimmed or changed. You must try and get the best out of your boat at all times and make the best use of the sea and the weather. You're in a race, whether you like it or not. The sea is impersonal, it doesn't give a damn about you or your boat and will punish any weakness that it finds.

As a single hander you cannot be on watch 24/7, so you have to have a system, that at least allows you some rest. I had several long voyages on the circumnavigation, and I found there were times of the day when things were going OK, and you could rest up for a while. No more than a couple of hours at a time. I could lay there in my bunk, no more than half asleep, listening to the noises of the boat making passage through the water. The sounds of water on the hull (a rustling sound), the noise of the wind in the rigging, halyards tapping on the mast, and a number of other little things, that tell me that everything is OK. But if something changes, and the noise changes, I'm up and onto it in a flash.

That system worked well in the trade wind zones, but how was it going to go in the Roaring 40s, where the seas were bigger and the winds more variable? The answer was that I got less rest—I found it hard to rest—I was like my companions, living in the moment, giving it my best, and prepared for anything.

We were about a week out from Cape Horn. It was just about time for the morning shot. I turned on the radio to WWV. I went to get my watch. I kept it in a very safe place, all wrapped up, still in the original box that I bought it in, with a few crystals to keep it dry. I put it on my wrist ready to check the time. The face was clouded over; I couldn't read the time. I tried to wipe the condensation off the face, no good, the condensation was on the inside. Wow, this could be a major problem. What to do. I started up the stove and held the watch at a safe distance above it. It didn't take long to dry out. I put it on my wrist and checked it against WWV. It was OK. I went ahead and did the sight. It all looked OK to me. I went below to update the log I left the watch on my wrist, maybe the warmth of my body would keep it dry.

It had been fairly rough while I was out on deck. The wind must have been at least 45 knots. 'Dorado' was fairly charging along; the swell was as massive as ever, but the tops were starting to develop breaks, and spray was flying everywhere. 'Dorado' was under severe strain. I knew I had to take another reef in the mainsail. Reefing the main in these conditions was heavy work, standing on top of the cabin wrestling the sail down, and tying in the reefs. I was nearly finished, just one more effort and we're there. I heard a loud click, then I saw the watch, sitting on my wrist. Oh shit… I made a wild grab at it… too late… the strap was broken, and it slid off my wrist, and onto the deck then into the scuppers. I tried to make another grab, but I was restrained by my harness. I probably would have gone over the side but for the harness. The seawater rushing down the scuppers took the watch, and it disappeared over the stern, and into 1000 fathoms of seawater.

I finished off the reefing of the mainsail and went below to

Chapter 6 - Cape Horn

check the level of the rum bottle. How could that have happened? I should have taken it off before I went on deck, but still, it was under a heavy longsleeved sweater, and my heavy-duty wet weather gear, which had a Velcro strap on the wrist. Then there was my sailing glove, which had a rubber wrist band. The fact was it was gone, lost through carelessness.

So, what were we going to do now? It had to be the ship's clock. Not ideal, but that was what we had. It kept good time and had a second hand which was quite easy to read. The minute hand wasn't so easy to read, and it would be easy to make a mistake there (an error of one minute of time means an error of 60nm in your navigation). Also, the clock was mounted on a bulkhead near the navigation table, which meant that there was going to be a lag between taking the sight and sticking my head through the hatch and reading the clock. Allowances can be made for these things, but there is always the potential for error. I've always kept the ship's clock on GMT as that is the basis for all navigation.

The weather now eased for a couple of days, and we had some variable winds, finishing up with a moderate nor'-east breeze which lasted for nearly 24 hours. A ridge of high pressure was passing through, and the barometer had risen. It wasn't going to last long, however, and the barometer started falling again. During that time, I got some good sights, and established a position that I was confident in.

We were about 500nm west of Cape Horn, and we'd be there in about three days at our current speed. We would cross from the deep ocean up onto the Cape Horn bank well to the north of Diego Ramirez (a small group of uninhabited islands approx.

50nm SW of Cape Horn) and pass to the south of Cape Horn, maybe even within sight of it. Our schedule of Cape Horn by Christmas was going to be met easily.

The weather stayed with us for a few more hours, and we had a glorious reach under full mainsail and genoa; like a Sunday afternoon sail. It was too good to last, of course, and by the next day the barometer was falling fast, and the wind had shifted to the nor-west and had increased to 40 knots. We were holding our course under much reduced sail. For the next 24 hours the wind increased, and shifted to west-nor'-west, then to west. The strength of the wind was outside my previous experience. I put it at 55-60 knots. The seas were massive, and everyone had a breaking wave on top of it. There was spray flying everywhere, and visibility was badly affected.

I had replaced the vane on top of the self-steering gear with a smaller one, more suited to these conditions. We seemed to hold our course. I'd taken the main off altogether, and all we had was a spitfire jib, and that was quite enough. We were having a 'Cape Horner', but how long was it going to last? I stayed below, anxiously peering out of the hatchway. The noise was unbelievable, and yet I thought I heard a muffled roar, somewhere astern of us. I had to stick my head out of the hatch to have a look. Some distance off on the port quarter was a huge area of white water. Where did that come from? I soon found out. Closer this time I saw what looked like a cross swell and where it met the main swell the water built up into a huge pyramid then came crashing down. We're not near any land or reefs, or any of these kinds of dangers.

We didn't have much time to think about it, because here comes another one. This time it approached us, and the boat was lifting over the swell, when it collapsed, and covered us in feet of

Chapter 6 - Cape Horn

boiling water. It pushed us over so that the mast was horizontal in the water, then the pressure came off, and we righted again. Unbelievable stuff. I just had time to glance around and check whether the mast was still standing when the next one was onto us. This time we went right over. I can distinctly remember walking around the hull until I was standing on the inside of the cabin top for several seconds, then the turn continued right round until we were left wallowing in the white boiling leftovers. A few squirts of water had come in around the edge of the hatch. That was all.

I was just about punch drunk, bracing myself for the next wave, but there were no more pyramids. Conditions had eased slightly (relatively speaking). The wind had shifted to the sou'-west and eased to about 40 knots. The seas were subsiding and threatened no immediate danger. Amazingly, the mast was still standing, and we were still moving forward. The spitfire jib was still set and pulling us along at four knots.

I got out on deck, and went round every piece of rigging, and checked everything that could be checked. It all looked OK. The self-steering gear seemed to be OK and was working away in its usual fashion. I went below to do some figuring. The sky was completely clouded over, so there'd be no sights at the moment, but one thing was for sure: all that turbulence we'd experienced was as we came in from the deep ocean, and up onto the Cape Horn bank. That meant we were about 50nm from the Cape, and we needed to reset our course to stay well clear of it.

The barometer had risen a fair bit after being very low, so we had the prospect of some reasonable weather, for a little while anyway. I went back on deck, and replaced the spitfire jib with the #3 jib, and two reefs out of the mainsail, the weather seemed to be easing slowly, and the sea conditions were a lot more friendly. The sky looked like it was clearing, but we were going through that period of the day when the sun was very low in the sky, and sights were not possible.

I went below and warmed up some soup. Warm soup and hard tack biscuits, hardly a feast, but boy did I feel better. I was sitting on the edge of the bunk, starting to relax for the first time in many days. A great weariness came over me. I wanted to sleep... not now, when I'm just about to make my greatest achievement. Just a little while, like we do in the trades. This is not the trades...

Something is wrong... something is very wrong. 'Dorado' is trying to tell me something... she's heaving around, like she was in a washing machine. She throws me off the bunk onto the cabin floor... I'm wide awake now...What's going on?... I can see out of a port... cliffs, covered in green moss, not 50m away. Oh My God!

I rush up to the hatch, and slide it open... I climb up on the engine cover, and stick my head, and shoulders out of the hatch... What can I do? There is a wave coming from astern... it's picking us up... 'Dorado' is surfing down this wave straight at the cliff... SSSSMACK!!! Into the cliff!

In that instant, everything disintegrated. The mast, the rigging, the sails, disappeared over the side... There was stuff flying everywhere... Something hit me on the head... I was blacking out, I was falling, falling... There was a small voice... It was saying, 'No, no... Not now... Go back, go back...'

I came to, and discovered I'd fallen back into the cabin. As I climbed back through the hatch, I found a scene of utter desolation... The deck had been swept clean; all those fittings designed to withstand many tons of stress just ripped away. There was already water above the floorboards. 'Dorado' was finished, and so was I, probably. The undertow was dragging us

Chapter 6 - Cape Horn

back... Another wave was coming, it was picking us up, but this time 'Dorado' broached across the front of the wave (a bit like a surfboard). We were moving across the front of the cliff past a place where the cliff had broken away, and now the cliff was overhanging, and we were coming up to a big rock... CRUNCH!! 'Dorado' was jammed between the overhanging cliff and the rock.

It was a precarious position, and it probably wasn't going to last very long. I had to go. I didn't know where we were, but if I stayed here, I was a dead man. At least get up this cliff somehow, then see where we are and what we can do about it.

I was in deep shock, but I was focussed, in a way I have never been before. I was going to try, and if it ended badly, then too bad! I was no worse off than I am here.

I went below for the last time. My personal papers, and the boat's papers I kept in a bag under the chart table, I'd take these with me anyway, just in case.

The water was up to my waist as I stood at the chart table trying to find my papers. Everything was under water. I pulled out a green bag... Was that it? Yes! I could see a passport in there. As I stood there, I could feel 'Dorado' moving beneath me. Yeah, that'll do... I've gotta get outta here before she goes.

Back on deck, I'm looking at my options... Only one really... Back there where the cliff had broken away might be my best chance... I'll have to get in the water and work my way along the cliff. I looked out to sea. it looked relatively calm. Waves come in sets, as every seaman knows, and right now was about the best chance that I was going to get. I'm in the water, fully clothed, including wet weather gear. The water is deep. I'm working my way along the cliff... it's not very far, but something is pulling at my legs... It's kelp... I can see it everywhere on the rocks. I get to the edge of the area where the cliff had broken away. It's all big boulders, covered in kelp... How am I going to get up there? The next set of waves arrives... I hadn't noticed them building up. The first one got me, and we went surging up over the boulders... We're close to the cliff... there's a big boulder coming up... there's a gap... we're going through the gap, AHHHH!!! I'm stuck... the pain in my ribs was excruciating.

The wave receded, leaving me scrabbling around. It was all kelp around me, and I couldn't get any traction. The second wave arrived... it was surging up around me... the pressure was building up... pop, like a cork out of a bottle... we're through.

The wave left me stranded near the high-water mark... a couple of steps, and I was on dry ground. What now?

My ribs were giving me plenty, but I could move, so they probably weren't broken. I had to get up to the top of this cliff and see what we could see from there.

It looked like a steep scree slope nearly all the way up. I remembered going up scree slopes on the Eildon Hills when I was young, three steps forward, two steps back. It was going to take some time. But before I start, I want one last look... I manoeuvre myself around, until I can see around the corner... There she is, sitting there, waiting for a wave to come along and dislodge her, and she'll sink into deep water, surrounded by kelp forever.

It was a pathetic moment. This didn't have to happen. It was all my fault. Utter carelessness, right at the point of achieving something good. We'd been together for 12 years, 'Dorado' and me, and we'd done a lot, and we had a lot more to do. Now this.

I wasn't out of this by any means, but if I do get out of it, I'm going to get another boat, and do the job properly next time. It

Chapter 6 - Cape Horn

was an emotional moment as I turned away and set about getting up this scree slope. Time passed; progress was slow, just keep going, step after step, head down, focus on the job. I stopped and glanced back, we were probably about halfway, and now I could see that the scree slope finished short of the top, and there was maybe 10 metres of cliff to climb. I also noticed that, on my right, the scree had given way to a grassy slope which seemed to veer off and rise steadily to the top with no cliffs. This might be a better option... Let's give this a try. On we go, step after step after... Whoa!... One more step and I'd have been gone. A huge chunk of rock had broken away, and had fallen into the ocean, hundreds of feet below. The chasm must have been at least 20 metres across. There was no way around it. I had to retreat to the scree slope and continue to struggle upwards.

Eventually, I got to the bottom of the cliff, and stopped to rest, to study how I was going to get up there. The cliff wasn't sheer. There were hand holds and footholds I could use. I just had to decide which route. I wasn't in great shape physically, and I still had all my sailing gear on; and I was carrying the green sail bag carrying my papers. I didn't have any choice, really. I just had to go for it. It took about 10-15 minutes. I scrambled over the top, and I could see... a lighthouse. There was only one lighthouse in this part of the world, and that was on Cape Horn Island. I'd been shipwrecked right on Cape Horn! There was a poetry in that, an irony, but I didn't think of that at the time.

Now that I knew where I was, my chances of survival had improved immensely. The sailing direction books that pilots use in this area mention the fact that the lighthouse has not been operating for some time. It also mentions that there is a fisherman's hut on the island, and fishermen are often there in the summer months. That was going to be my destination. One other fact from the pilot book: it said the highest point on the island was 1500ft above sea level. The high point wasn't far from where I was, and it was easy to estimate that if the high point is 1500ft, then where I came up must be approx. 900-1000ft. Looking across from where I was standing, I could see a hut, it looked like about 3-4km in a straight line.

Could I travel in a straight line? There was no path, just a steep slope of what looked like heavy, tussocky grass. I took one step into the grass... My left leg disappeared, right up to the thigh. I fell over on the steep slope, and pain shot through my left leg. I somehow struggled back to my feet. My left leg was hurting, quite badly, but nothing was broken... A bit of movement would be good for it, probably. Straight across was not an option anymore. I was going to have to stick to the edge of the cliff, all the way down to the lighthouse, then down to the beach front, where there were two small bays separated by a rocky outcrop. After that, we'd see what it looked like.

There was nothing at the lighthouse. It had been built by the British, and the date and the names of the builders was all the information I could find. Nothing for shipwrecked sailors. Just before I moved on, I thought, I've been lugging this green bag around all day... Let's have a look see what's in here. I'd put in a couple of lumps of cheese, I knew that, and also some dry biscuits. Then we come to the paperwork... Let's see what's here... The passport, I'd noticed, was there. It was an old one, out of date, cancelled. Something wrong here... Where are my current papers? I pulled the rest of the papers out of the bag, and started going through them... They were old taxation papers going back years. Unbelievable, I had deliberately collected all the important papers and stowed them with the life raft, in a soft pack, these other papers should have been thrown out years ago.

Chapter 6 - Cape Horn

How they survived, and ended up where they were, I couldn't imagine. It just means that the only ID that I had was the out-of-date passport. But I'm not out of it yet, so we don't need to worry about it for the time being.

The next target were the two small bays. I was about halfway around the first one when something caught my eye. There was movement on the high ground ahead of me, it looked like four or five people moving in single file down a track coming my way. The spirits jump

immediately... I'm waving my arms and jumping around and shouting Over here! Over here! They disappeared from view. Did I really see something or was I imagining it? Then they began to reappear lower down the hill. I stared in disbelief... they were penguins, bloody penguins. Oh dear, settle down, get a grip, carry on.

Around a bit of a rocky point, and into the second bay. This bay is a bit bigger, and about halfway along I could see a stream running out of the thick bush and across the beach. I hadn't brought any water with me, and I was thirsty. The stream water would surely be drinkable. As I neared the stream, I could see penguins gathered around it. All their heads were turned my way, watching me carefully. They didn't seem to be scared of me, and they didn't run away. I found a suitable place then washed my hands. It looked OK, so I cupped my hands and drank some water. It tasted OK. I had another drink. I needed that. I was about to have a third go when I noticed one of the penguins lift his leg, and a stream of urine shot into the stream. Oh yes, thanks very much, maybe there'll be another stream farther along.

I was getting towards the end of the bay, and it was becoming obvious that I was going to have to change course. My destination, the fisherman's hut, was straight ahead of me, but that meant climbing another cliff, and then up and over the hill. There had to be an easier way. I'd seen the penguins come down the side of the hill… There must be a path. To go up and around the hill was certainly going to be longer, but it would be easier and with less risk. I found the path easily and set off.

As I moved up and away from the beach, the thick bush subsided, and the wind became more of a problem. It was strong, and it was cold. Despite all the clothes I was wearing, the cold was eating away at me. It was hard to keep going. I was progressing slowly, up and around the hill. If I could keep going a bit longer, the wind would be behind me. The going was a bit easier now, just a few gullies here and there. The sun was sinking below the horizon. At this latitude it is only below the horizon for about half an hour, and there is always a glow, so it is never completely dark.

I saw the gully as I approached it. Most of them, I'd just walked through, but this was a big one. I realised too late, and tried to stop and go round it, but I stumbled, and fell right in. It was full of thick bush which caught me and cushioned me. It was beautiful, the softest bed I've ever been in. I just want to stay there for ever, for ever. The alarm bells were ringing… You gotta get outta here, get moving.

It took ages to struggle out of there, and I was lucky that I had all my heavy-duty wet weather gear on, otherwise I would have been badly scratched. Get moving again. Get motivated. Where the bloody hell is this hut anyway… It couldn't be too far now. One more rise, maybe I'll see something this time. Yes, there it is, that's it… and there's a light in the window! I stumble on, not really looking where I'm going. Suddenly, I trip over something and fall flat on my face. I'm paying attention now as I struggle to my feet. Barbed wire, barbed wire everywhere. What's

Chapter 6 - Cape Horn

this, a bloody minefield? If it had been a minefield, I would have been mincemeat by now. What can I do? Pick my way out of it carefully… The light is improving…

I'm approaching the hut now, round the side where the light is in the window… I get to the window, and look in… There is a room with a large table, and chairs, and a young fellow in military uniform. He has his back to me. He is studying a chart laid out on the table. I lift my hand and tap on the window… Tap, Tap, Tap! He spins around, and he stares in shock at what he can see outside the window. He backs away and disappears from my sight. A few seconds later he comes round the side of the hut, with a dog on a lead, it's a German shepherd, and it's straining on the lead.

I needn't have worried. The dog had smelled the dried blood on my wet weather gear, and that's what it was interested in. The guy spoke to me in Spanish. I didn't have a clue what he said. I said, 'Por favor, senor, no hablo espanol'.

By that time, several other young guys had appeared on the scene. They took me inside, and sat me down at the table, and we started to try and communicate with each other. They were a platoon of Chilean marines stationed there to keep an eye on the Argentinians across the Beagle Channel. This was 1983, not long after the Falklands war, and the Argies are still a bit trigger happy. The Chilean navy has very close ties to the British navy, and the Argies resent that.

One or two of them had a smattering of English, and we soon worked out that I was a yachtsman and had come to grief on the cliffs. It felt so good to be back in the land of the living. I think I was a bit overcome by it all. They got me a coffee, and something to eat, then somebody stitched up the gash on my forehead, then somebody came up with a bottle of whisky. That was too much.

In the meantime, the officer in charge had been in contact with his base at Puerto Williams, about 120kms to the north as the crow flies. A supply boat was due in the next day or so, and I could go out on that. Even though it was daylight the local time for them was about 3am, and they wanted to get back to bed. They wanted me to go to bed, as well. They gave me a bunk in a room with all the blinds down, with some of the marines. The bunk was hard, and there was no chance of any sleep. I was in pain. My ribs were giving me plenty, and every other muscle in my body was screaming out. I tried to get up, but I couldn't. I needed some painkillers, but there were none around.

The next thing I remember was the chattering of a helicopter. The marines had decided that I wasn't in very good shape, and that I needed to be checked out in hospital, so they sent for the helicopter which took me up to the naval base on the Beagle Channel, not far from Puerto Williams. The hospital x-rayed my ribs. Nothing broken, but badly bruised, and would be painful for some time. The rest of me was essentially OK; just banged up a bit.

Mentally, I was still in shock, and suffering from exposure. They took away all my clothes and left me in a dressing gown. They put me up in the sergeants' mess and told me I was to be interviewed the next day. I got my clothes back the next morning, and soon after I was interviewed by two Chilean Navy intelligence officers. They were very correct, and spoke reasonable English, and seemed to be mainly interested in whether I was a British agent escaping from Argentina. They seemed a bit disappointed when I told them, no. The interview didn't last long after that.

Then a lady entered the room. She introduced herself. She was English, married to a Chilean Navy Officer. Her beautiful

Chapter 6 - Cape Horn

English accent charmed me completely, and we spoke for nearly an hour. She got the whole story, and we also spoke about what was going to happen now. Tomorrow, I would fly to Punta Arenas where I would be met by the British Vice Consul, and my ongoing travel would be arranged by the consulate.

After the interviews, I was at a loose end for a few hours, and I thought I would go for a walk. I'd finally gotten some pain killers and was feeling much better. The weather was about as nice as you could expect in that part of the world, as I set off down an unsealed road which followed the coastline. I hadn't gone far when my thoughts started to turn to my predicament. What a mess I'd made of things. I'd let myself down, and I'd let 'Dorado' down. I'd just abandoned her to save my own skin. She was going to stay there for ever at the bottom of those cliffs. There was a lot of my personal gear there too: my camera, and hundreds of photos, logs and records of previous voyages. I'd pretty well made up my mind that I was going to go back to 'Dorado' when I heard a vehicle coming down the road behind me. It stopped alongside me, and the driver spoke to me.

'George,' he said, 'where are you going?'

'I'm, I'm just having a bit of a walk,' I said, suddenly returning to reality.

'You've gone a fair distance,' he said. 'Would you like a lift back?

Yes, I said, thanks very much.

The pain killers were wearing off, and the aches and pains were returning, and there was something else. My mental health was a bit shaky. That was something I was going to have to watch. Next day, I flew up to Punta Arenas on a Navy transport plane, a Hercules C130. There were other people on the plane, and a bit of cargo. Seating was very basic. The noise inside was unbelievable. Those aircraft are not lined, so you get all the engine noise inside the cabin. It's possible to converse generally only through radio between passengers.

Conversation was out of the question.

On arrival in Punta Arenas, I was met by a tall, well-dressed, and well-spoken gentleman, the British Vice Consul. He told me he was a sheep farmer, and his family had migrated to Chile from the Scottish Borders generations ago. He also told me the British authorities had been in touch with my brother in Scotland, and he had put up the money to fly me back to Scotland. The alternative was to wait around in Punta Arenas until we could find some cargo vessel returning to the UK that could offer me a berth.

I was going to fly, thanks very much. I had no money on me, of course, but that could be sorted out later, no problem.

We had a meal, and he asked me whether I would mind meeting the media. I was interviewed for TV with a translator. I saw some of it later on TV. Disgusting. I looked a sight with a great big plaster stuck on my forehead.

Next day, I flew from Punta Arenas to Santiago where, again, I was met by a British representative. I didn't have a passport, of course, so I was going to be travelling on a special once-only document. Passports could be fixed up after I got to the UK. I was to fly from Santiago to Rio de Janeiro, where I would change planes, then fly to Gatwick, where I would board a helicopter to Heathrow just in time to connect with a shuttle flight to Edinburgh. All of this happened without any hiccups, and my brother was there to meet me in Edinburgh. I'd phoned him from Santiago. He already knew of my disaster, but I just wanted to assure him that I wasn't too bad, and to prepare Mum for my arrival.

Chapter 6 - Cape Horn

I was home in the borders just before Christmas, and of course it was an emotional time, and also a time to recover.

Chapter 7
Consequence

My ribs took a long time to come good. They were still paining when I flew back to Sydney in mid-February, but what worried me more were the nightmares I began to experience, which continued every night for quite some time. They stopped eventually, and I have never dreamt again to this day. The Australian Government has an immigration office in Edinburgh, and in mid-January I went to see them about getting a new passport. They were very helpful; they organised a new passport for me within a few days. (I am a dual national and entitled to hold both).

That cleared the way for me to fly back to Sydney in mid-February, after the summer holiday rush was over. I had no home to go to, of course, but my good friend, Joe Grech, had offered to put me up for a few weeks, until I found my feet. Getting back to work, and getting some money coming in, was the priority, and that is exactly what I did. Driving in Sydney traffic, finding a fare, and dealing with the Sydney travelling public is an all-consuming exercise. It doesn't leave any time for negative thoughts, and that was a good thing for me right then.

I had a plan, of course. I was going to get another boat, and sail it round Cape Horn, properly this time. But first I had to work for a few years and set myself up financially.

By the time that I'd done three years, I was starting to taper off. I'd decided to build a new boat from scratch, and I'd already been to see the designer, and had the plans to build an Adams 35. Whilst I was there, I asked him if he could recommend any builders. He gave me a list of people who were building his designs. I checked all of them, but they were all busy, and they

Chapter 7 - Consequence

were building the new 40' design, and not interested in the older 35' design that I wanted. One of them said to me, you're a Balmain boy, maybe you know Jim Smith, he's a Balmain boy, and he's building a steel boat over at Riverwood. Maybe he could help you out. He gave me the address, and I went over there, and found them behind a house on Salt Pan Creek. I recognised Jim's face straight away; he'd been a worker at 'Browns Boatbuilders' next to Conway's at Balmain when I was there. We chatted for a while, and he was interested. We came to an agreement and went from there. He found a site at Taren Point, on an oyster farm There were 10-15 people there, building boats, and we were quite a little community. I kept working, but the number of shifts declined as the work progressed. I just wanted to be involved in the building, and after a while I was working only at weekends.

I'd been living in a flat at Ashfield for the last three years, but I was hardly ever there. I was down at Taren Point five days a week working on the boat and driving a cab on Saturdays and Sundays, so I wasn't doing much more than sleeping there. It wasn't going to be very long before I'd be living on the boat at Taren Point. One day I was just about to leave the flat when the phone rang. Who's the lucky person that got me today. Hello! It was Colleen Manton, my old boss's wife, back in Sydney. I knew that they had sold up in Sydney and gone up the country before I'd left in 1982. I'd also heard rumours in the cab game that they'd come back. It's true she said. Joe left me and got shacked up with my best friend. I'm staying with Shelley at Gladesville at the moment. I've had a disaster too I said, lucky to be alive really, but I'm recovering, and getting ready to have another go. Good for you she said, but listen, it's my birthday shortly and we're having a small party, would you like to come and join us, it's on Saturday night. Yeah, why not, I haven't done much socialising recently,

I'll see you on Saturday.

Saturday night, and I'm getting ready to go out, trying to make myself at least semi-respectable and thinking, are you sure you know what you're doing here? Yeah whatever… where's that bottle of wine? And, of course, when you least expect it, it happened, and Colleen, who became Chloe, and I got together and became an item, and we're still together 35yrs later. She is the love of my life!

The big test came when we moved out of Ashfield and down to Taren Point. The boat was at a stage where it was nearly ready for occupation as our new home. Chloe had a job as a telephonist with Telecom, as they were known then, and I was busy with the boat, and still driving a taxi on the weekends. The move from Ashfield to Taren Point was a big one, especially for Chloe. Down there at Taren Point, on the oyster farm, among the mangroves on Woolooware Bay… where the discarded oyster shells gave off a pong on the hot days, became flooded when it rained, and was thick with mosquitos at night. It was hard to believe we were living in the Sydney metropolitan area. There were several other women there, however, even kids, so we were quite a community. There was a shower, which needed a coin or two to give you a hot shower, and there were regular barbecues.

There was another advantage of being in a community — whenever you needed something, a piece of gear, a service or anything, somebody knew how or where to get it, and at the right price.

The boat was advancing apace — much quicker than most of the other boats at the 'Swamp'. Some of them had been there for

Chapter 7 - Consequence

years. One resident, Andy Frew, reckoned he'd been there for 15 years. My new boat was to be named 'Consequence'. The steelwork was finished, the lead was in the keel, the auxiliary motor was installed, the steel had been painted with three coats of tar epoxy outside and in, and the interior fitout had been completed to the stage where Chloe and I could live on board. I'd planned to have the boat finished launched and sailing inside two years, and we were on track to do that. The mast and rigging, was a big item to come, a suite of sails, self-steering gear (an Aries, same as 'Dorado') and electronics.

'Consequence' was launched in early 1988, and we had a mooring in

Woolooware Bay. We did our early sailing trials in Botany Bay, and later moved up to Sydney Harbour and back to Balmain. (Jeremy Lawrence still owned the Balmain Marina, but he had leased the business and the manager was a very nice guy indeed). Our first serious voyage was going to be to Lord Howe Island, but it never happened. We went through Sydney Heads on a day when there was a fresh sou'-east wind blowing, and a leftover sea still running from a strong southerly a few days earlier. It was very bumpy, and it wasn't long before we had our first problem. The life raft was secured on deck, in its cradle and strapped down accordingly. A rogue wave came along and washed over the foredeck. It seemed to lift the life raft out of its cradle and hurl it to the port side. It would have gone right over the side except that the sail caught it and dropped it down in the scuppers. I grabbed it before it disappeared over the side, and put it back in its cradle, and secured it. This was a problem that was going to have to be looked at again.

Another, more serious problem emerged soon after. The compass was playing up; when the boat heeled over, it was going haywire. Setting up a compass on a steel boat is tricky, and I always had it done by a professional. Captain Bozier had set up the compass on 'Dorado' and adjusted it three or four times over the years, and I was very happy with him. I got him again to set up 'Consequence', and straight away he didn't like the position of the compass.

'It should be higher up and further aft,' he said. 'It should really be set up on a pedestal. But I'm here now so I'll do my best with it where it is.'

He was right, of course. It needed a pedestal, and it was out of the question continuing with this trip. We were off Barrenjoey at this point, and so we just made a fair breeze of it into Pittwater. We spent a few days up the Hawkesbury, then returned to Balmain.

'Consequence' was still brand new, and small problems could be expected to show up. I'd also noticed some vibration around the rudder stock. Some stiffening might be required around there.

We never got to Lord Howe on the boat, but we did visit there four times, flying in as tourists. It remains just about my favourite place to visit. I still wanted to go back and take on Cape Horn, but Chloe wasn't up for that. We talked about a voyage round Australia… Yes, she was interested in that. That would be a suitable challenge for both of us and would be a good lead up for me to take on Cape Horn.

We left Balmain in April 1989 and headed north up the NSW coast. Our first stop was to be Coffs Harbour which has a modern

Chapter 7 - Consequence

marina with all facilities. Chloe had a number of family members living in the area, so we'd be doing a bit of socialising. The weather was a bit mixed on the way up, mostly light to moderate winds, a bit cloudy, and some rain. Coffs is a nice place to stay, and we enjoyed it there.

Our next stop was Mooloolaba on Queensland's Sunshine Coast. It was about a week's voyage from Coffs, and again it was mostly light weather, until we got to the northern end of Moreton Island. As we started to round the cape, the wind began to increase from the nor'-west and was soon gusting up over 30 knots. Mooloolaba was about 40nm to the nor'-west of us, so it was going to be an interesting night. There was no ocean swell to speak of, just short, steep seas, every swell with a crest, so it was bumpy, and it was wet.

Chloe stayed below whilst I stayed on deck. I was tacking regularly, and there was a chance of traffic coming out of Moreton Bay.

As the night wore on the breeze eased a bit, and by dawn we were in sight of land, and sailing fast. We were a little south of Mooloolaba, so, one more tack, and zoom, we're entering the creek. I'd stopped at Mooloolaba in 1977 at the start of my earlier circumnavigation. I'd found it a very handy place to stop, and the marina welcoming. This time it wasn't so friendly, although we did get a berth for a week.

Our next destination was Fraser Island, reputedly the largest sand island in the world. I had previously sailed up the eastern shore, but this time we were going to go up the western side, up the Great Sandy Straits, where there is a well-marked channel. The access to this channel is across the Wide Bay Bar. It's not all that difficult, but it's not lit, so you must do it in daylight hours. It's also recommended that you do it on a rising tide. To satisfy all these criteria, we had to leave Mooloolaba at dusk, and travel north overnight to arrive at the bar in daylight and on a rising tide.

All went well, and we arrived after dawn, and picked up the leads, and proceeded to cross the bar. Just at the critical point of the crossing, the auxiliary conked out. Panic stations! I rushed around, hoisting some sail, and we sailed/drifted with the tide to a safe area, where I dropped the anchor. The problem with the auxiliary was the same as I'd had with 'Dorado'. Some small piece of detritus had gotten into the water pump and blocked it; the engine overheated and stalled. I had it fixed in 10 minutes, and we proceeded around Inskip Point to an anchorage close to a settlement.

It was a busy place. There is a barge that services Fraser Island, and it's very popular with four-wheel drivers; they just love to race up and down that lovely white beach that goes for something like 120kms. There is vegetation, almost a jungle, but it sits on top of the sand, and is very easily destroyed. There are also dingoes, and they are not friendly. There have been incidents. The whole place should be declared a national park, and the public banned, especially the 4WDs. Sandy Strait was ahead of us, but if we looked the other way the straits continued for a short distance, and ended up at a place called Tin Can, on the shore of Tin Can Bay. What a ridiculous name, how could any town be called Tin Can!

We just had to go and have a look. So next day, we took off down the narrow channel that took us there. Tin Can is a very nice little town, and it relies on tourism. There is a marina, which we stayed at, and had a few pleasant days. The name derives from an Aboriginal name for some feature, which existed at some time in the distant past. I suppose if you depend on tourism for

Chapter 7 - Consequence

your livelihood, then anything that attracts attention is good. It attracted us, didn't it!

From where we were, it was almost 100nm through the Great Sandy Strait. We were going to take two days to do it, not travelling at night in such restricted waters. On the first day, we made it well up the main channel, which is fairly wide, and we were just about to anchor for the night when we noticed another yacht. It must have wandered off course into shallow water, where it had grounded. There were people waving to us, and they got into a dinghy and came over to see us. They turned out to be people I'd met in Fiji years ago. We had a meal and a couple of drinks and told a few stories. The tide was starting to rise, and they had to get back to their boat as they were hoping to get off before high tide. When we woke next morning, they were gone, and they were headed in the opposite direction to us, so we never saw them again.

Our plan was to get through the straits, nearly to the other end, then anchor for the night. The centre of the straits featured narrow, but wellmarked channels. We kissed the sandy bottom a few times, but other than that had no problems. We found a nice anchorage near the entrance to Wide Bay.

I'm not much of a fisherman, but I try my hand occasionally. Coming up the coast of NSW, we'd troll a line, and a couple of times we'd caught a mackerel, which was nice eating, especially fresh. This time I just had a handline, which I threw out from the boat. I got a couple of nibbles, and then bingo, something bit, and I hooked it and hauled it in. It didn't look right; didn't look like the kind of fish that you eat. As soon as the Dear Lady saw it, she pounced.

'That's a Puffer fish,' she said. 'It's poisonous! Throw it

back.'

Early next morning, we cleared out of the straits, and set off across Wide Bay for the Burnett River. There is a safe anchorage near the mouth, and that's where we were going. We were most of the way there when we came up behind another boat, and as we neared it, we could see that it was one of the boats from Taren Point. It was John, one of our friends. We pulled alongside as close as was safe and had a conversation with him. His engine had broken down, and he was trying to make Burnett Heads before dark. The breeze was very light, and he just wasn't going to make it. We offered him a tow. It did slow us down of course, but we made it, just before dark.

We rafted him up alongside us for the night, had a meal and heard his story. He'd been in Mooloolaba, he said, although I don't remember seeing him there. He'd filled up his fuel tank there before setting off, and he thought he'd probably taken some dirty fuel. When the engine broke down, he was much closer to Bundaberg than anywhere else, and was trying to get there. In the morning we offered him a lift up to

Bundaberg, which was where we were going, but he said, no, he'd stay with his boat, and try to get the engine going. There was no hope of towing him up the river against the current, so we wished him well and set off up the river.

Bundaberg is a large town/small city, the centre of a sugar cane growing area. There is a sugar refinery in the middle of town, and right next door there is a distillery which makes the famous 'Bundy Rum'. I'm quite partial to a 'Bundy' or two on sailing trips. It seems to sit better with me than other spirits. We got ourselves a fore and aft pile mooring at Bundaberg right in the middle of town, just near the bridge. It was a great place to be and gave us easy access to the middle of town. The weather was

Chapter 7 - Consequence

great, as it should be at this time of year, and we very much enjoyed our stay there.

Too soon, it's time to go, and as we make our way down the Burnett River (much quicker this time, thanks to the current being behind us) our thoughts turned to John. We hadn't seen him in Bundaberg. What was he doing? When we got to Burnett Heads, we could see his boat at anchor. We went over to check him out, but he wasn't there. Later, on the grapevine, we heard that he'd gotten up to Bundaberg and ended up staying there.

Our next stop was to be Port Clinton, on the edge of the Shoalwater Bay military reserve. We were passing inside the bottom of the Great Barrier Reef by now, past Lady Elliott Island and Lady Musgrave Island, popular with yachties with its enormous, quiet lagoon, and on past

Heron Island and the Capricornia Cays. Port Clinton, near Rockhampton, was a nice secure, anchorage, but there was no habitation around, just mangroves, mangroves, mangroves. Not even a place to land.

Next day, we set off for the Percy Island group, but the weather was very light, and we ended up anchoring off the northern side of the military reserve. It was a poor anchorage, and very rolly. We wanted to go ashore and stretch our legs, but through the binoculars I could see a big sign erected on the beach: 'No Entry, No Landing, Military Exercises taking place. Live ammunition being used!' That was that.

Next day, we set off for Middle Percy. This Island is famous because of the people who live there, and the hospitality they show to all. We were most of the way there, and I was looking through the binoculars. I could see a number of yachts there, all in motion, rolling, etc. It's known to be a bit rolly at times and that would be two uncomfortable nights in a row. We made a snap decision. From where we are, we are closer to South Percy, so let's go there.

We anchored off a sandy beach for the night. There was a bit of movement, but it wasn't too bad. We had the whole place to ourselves. We went ashore next morning, had a good walk and, feeling refreshed, we decided to sail over to Middle Percy to anchor for the night. There were several boats there, and the anchorage was uncomfortable. We didn't go ashore and left early the next morning. Did we miss something? I doubt it. With the number of boats there and the number of people, hospitality might have been a bit strained.

Our next anchorage was to be at Thomas Island, at the southern end of the Whitsunday Island group. Alan Lucas, in his book 'Cruising the Coral Coast', gives it a good write up: beautiful, safe, secure anchorage, he says, and it was. The breeze was up, and we made a fast passage, arriving late afternoon. We had a peaceful night for a change, and in the morning, we went ashore to explore.

We had a number of choices for our next stop, and we settled on Hamilton Island which is a holiday resort with large tourist hotel, and a landing strip that takes jets direct from Sydney. The whole place is very commercial, and expensive, but we needed a bit of a break after a series of pretty ordinary anchorages. There is a large marina, and without any trouble, we found a berth for a few days. Chloe is not as seasoned a yachtie as I am, and she really appreciated having a hot shower every day, and getting the washing done. We didn't meet anyone we knew at Hamilton Island, which was not surprising, and were glad to get away from it in the end.

Our plan was to visit a few more anchorages around the

Chapter 7 - Consequence

Whitsundays, finishing up in Butterfly Bay, then direct from there to Townsville. It is about 135nm to Townsville, so the sensible thing was to leave in the morning and travel overnight, arriving in Townsville sometime after lunch. The trade wind was well established, and we had a fast passage, arriving in Townsville before lunch the next day. We went right into Ross Creek and found a fore and aft pile mooring without any trouble.

Amazingly, one of our near neighbours was the yacht 'Impossible' from Taren Point, with Jimmy and Karen. Based now at Balmain, we'd gotten a bit out of touch with Taren Point. We'd known they were just about to launch when we left Taren Point, and here they were. So, there was a bit of a reunion, and stories were swapped, and beer was consumed.

Chloe also knew some people in Townsville, and she hadn't heard from them for ages. She called them, and we were invited out. The husband was a sales representative and travelled all over North Queensland. I mentioned that I'd worked at Mount Isa, and at Cloncurry 20 years previously, and names were mentioned, including my nemesis, the detective sergeant from Mount Isa. He laughed at that. The guy was notorious, he said. He'd been moved around from place to place, but caused trouble wherever he went, and in the end he'd been forced out of the police force.

Townsville is a city of 150,000-plus and has every facility that the average person might need. It would be easy to stay here among friends and get stuck. We just about had to force ourselves to leave, and when we did, we only went as far as Magnetic Island, 8kms offshore. There, we were able to go ashore to shops, and a pub, but the local people seemed to have a rather negative attitude, and we stayed only for a couple of days.

Our next destination was Hinchinbrook Island, another

110kms north. We were headed for the channel that separates it from the mainland. But before that, we wanted to explore an anchorage on the eastern (seaward) side of the island. There's no official chart of this small area, but it's well described by Alan Lucas. There's a small bay, Zoë Bay, easily identified, and at the northern end of the bay there is a bar leading into a creek, which is a calm and secluded anchorage. In the wet season, the water must pour down here off the high ground, and rush over the bar. The bar changes continually, which is the nature of these things.

We timed our departure from Magnetic Island so that we would arrive in the bay at high tide, and we did, on a morning with a light breeze, and no running sea, which was ideal. We got to the north of the bay, close to the shore, and followed the darker water. We were about halfway through when we stopped. We'd run aground. It was only soft sand, and it was easy to get off. Try again, a little further over this time. Same result. This time we had to back right off. Back out into the bay, I dropped the anchor. It was good holding ground. We launched the dinghy and headed over to the bar and proceeded to sound our way through. We went back and forward for 30 minutes, but the best we could do was a depth of 1.3m. 'Consequence' needed 1.6m, so it was obvious we weren't going to get in there today, which was a disappointment. But we did go in there in the dinghy to check it out. It was everything that Lucas said it was, a great anchorage in more than two metres of water, beautiful surroundings, fresh water cascading down the hill, but not for us, not today. As I write this in 2021, I am looking at it on google earth, and I can see the bar quite clearly, and it looks as if it would be navigable, especially at high tide. There you go.

It had been a tiring day, and we decided to remain at anchor for the night. The wind remained moderate, and there were no

Chapter 7 - Consequence

problems. In the morning, we hauled up the anchor and set sail for the northern end of the Hinchinbrook channel. The area is home to the dugong, a threatened species which lives on the seagrass found in that area. There was talk at one time of building a large tourist resort at Cardwell, which would have interfered with the dugongs, but there was a loud protest movement, and the plan didn't go ahead. We anchored near the nor'west corner of the island, where we could see a barbecue area, so we went ashore to investigate. The first thing we saw was a big warning sign regarding crocodiles. 'Do not attempt to stay overnight. Get out an hour before dark.' This was already 4pm on a winter's afternoon, and the place had a strange feel about it, as though you were being watched. We beat a retreat back on board and settled down for the night.

Next day, we explored the channel a bit more, and some of the side channels. It was instructive to see the huge crocodile slides on the mudflats. We did see one or two smaller crocs, but there are obviously plenty of big ones around. Crocs have been protected in Australia for many years, but they are farmed in some places for their skins and their meat.

Our next move was going to be to Dunk Island. The anchorage is not very good, but the place has a very well-known resort, and makes visiting yachts welcome. On the way, we passed Bedarra Island, which is quite exclusive, and doesn't welcome visiting yachties. The whole trip is just a day sail with a fresh trade wind, making sure we wave to the people on Bedarra as we go by.

Dunk Island is certainly a beautiful place. It has an airstrip; planes fly in and out every day. Yachties are allowed to use the facilities, which is great. Shame about the lousy anchorage.

The town of Mission Beach sits on the Queensland coast not

far from Dunk Island. Chloe had friends in Sydney (he was a cab driver), who had sold up in Sydney and moved up to Mission Beach. They had a business, transporting tourists to Dunk and Bedarra, on a hovercraft, no less. The morning after we arrived, we saw the hovercraft coming, and went ashore to meet it. The hovercraft tied up to the jetty, and the passengers disembarked, then the owner/driver. He did quite a double take when he saw Chloe, but settled down fairly quickly, and offered to shout us a beer. 'I'm not going back for a couple of hours,' he said. 'We've got time for a couple of rounds.'

One or two rounds turned into three or four rounds, and all of a sudden it was time for him to go. He invited us to come with him, meet the family, and stay the night. Chloe was keen to go, but I wasn't so sure. The anchorage was iffy at best, and with a number of other boats around, I didn't want to leave 'Consequence' unattended overnight. So, Chloe went, and I stayed behind. Next day, the entire family came over, mum dad and four kids, and we had a slap-up lunch at one of the restaurants. He refused to let me pay, even our share. So, a pleasant time was had by all.

Next morning, we were a bit hungover. We decided we wouldn't move on today; we'd just have a lay day and move on tomorrow. We'd originally planned our next stop at Innisfail, up the Johnstone River. But a major cyclone had hit that area the previous year, wiping out all the navigational beacons, and the sandbanks had all shifted position. Navigation was just about impossible. So, the sensible thing was to go to Mourilyan, just short of the Johnstone River, which had a deep-water port and a dredged channel to allow access for the sugar ships.

It's only about 35kms from Dunk Island to Mourilyan Harbour, less than a day sail. The directions say keep well wide

Chapter 7 - Consequence

of the coast, until you pick up the leads, then proceed straight into the harbour. We came around the headland, and saw all the boats at anchor, and lo and behold, there was 'Impossible', they had gotten the same message that we had, and had come to Mourilyan instead of trying to get up the Johnstone River. The anchorage was good, and there would be no cyclones around at this time of year, so Mourilyan was a safe harbour. There was no habitation there, no shops, no civilisation, except for the sugar wharf. Jim and Karen were talking about getting a taxi into Innisfail. Were we interested? Yes, we were.

Innisfail is the centre of a rich agricultural area, mainly sugar and bananas, and is a sizeable country town. We spent a pleasant day there. There were yachts at anchor in the river, and we spoke to some of the yachties. It's the bar that was the problem, they told us. You can come over just about anywhere at or near high tide, and once you're in the river there's plenty of deep water. These people had also been to Zöe Bay on Hinchinbrook Island. We told them about our problems with the bar.

'Oh, the bar is OK,' they said. 'Just take a runner at it. The momentum will take you through.' Yeah, I hear what you say!

Mourilyan wasn't the place where you'd want to stay for any length of time. It was a good anchorage, but there was nothing else there. So, our thoughts turned to Cairns. Cairns is a small city, not quite as big as Townsville, and focussed more on the tourist industry. It has an international airport, and tourists come year-round, mainly from Asia. At Cairns, you're in the genuine tropics (even in Townsville we got the sou'-west wind with an edge to it, all the way from the Southern Ocean). Mourilyan-Cairns is about 50nm , and if you leave early in the morning with a good fresh trade you should be in Cairns before dark. And we were, anchoring in Trinity Inlet just opposite the yacht club. There is a marina just next to the yacht club, but we'd been advised not to worry about it. Next day, however, we went over and joined the yacht club, which gave us a place to land our dinghy, hot showers, washing facilities, and a nice bar and restaurant.

One of the things we wanted to do in Cairns was to visit the butterfly farm, up on the tablelands at Kuranda. There is a scenic railway which climbs up to Kuranda, with the town sitting almost on the edge of a cliff overlooking the Cairns coastal plain. The train runs in the dry season, as there are huge waterfalls in the wet season that make it impassable. It's certainly an impressive trip. In the dry, the waterfalls are only a trickle compared to the wet season. It was a very enjoyable day.

We were sitting in the bar at the yacht club one evening when two people approached us. It was Gerhardt and Susi. We had met them down in the Hinchinbrook Channel on their yacht, 'Heimdahl', now here they were in Cairns. Susi had a friend in Cairns who had a Kombi van, which she was lending to Susi for a couple of weeks. They were planning to head up onto the Atherton Tablelands to visit some of the lakes up there. Were we interested in going with them? Yes, we were! 'Impossible' turned up in Cairns the next day, and Jim and Karen came too, making six. What a pleasant day that was, away from the heat and humidity of Cairns. We all had a dip in Lake Barrine. It was cool to cold in the water; very refreshing. So, a great day was had by all.

We were anchored in Trinity Inlet, more or less opposite the yacht club, close to the mangroves, which stretched for miles to the south and west, and were crawling with crocodiles. People flying into Cairns' airport could see them and reported them.

Chapter 7 - Consequence

That meant absolutely no swimming, and no transfers in the dinghy at night. But there was a lady on a nearby yacht, 'Wings', who was doing some fishing at night out of a small dinghy. We found out one morning at breakfast, when she—her name was Kay—came around and knocked on the hull.

'I went fishing last night,' she said, 'and I've caught more than we need.

Would you like some?'

The upshot of this was that Chloe wanted to go fishing, too, so we went ashore and got her a prawn net, and she went prawning with Kay. They got prawns every time they tried. I was concerned about their safety, but nothing happened. The lady's husband, Peter, was confined to the boat. He'd hurt his back and couldn't get into the dinghy. It was a funny situation. There are doctors in Cairns, and an excellent hospital. If he needed help, this would be the place to get it. I thought of going over and offering to help, but Chloe said they'd already been ashore and seen a doctor and had some medication. Tough thing to happen on a cruising boat.

We'd had an interesting stay in Cairns, but it was time to move on. We'd thought of going over to Green Island, one of the very first upmarket tourist resorts to be set up on a coral island, but the locals at the yacht club advised us not to go there. You can't tie up at the pier, and the anchorage is very poor on coral, and then you're a long way from the shore. If you do all of that, and you do go ashore, the locals on the island won't welcome you. But if you 'dead set want to go out there', there were ferry trips from the wharf every day. That was the way to see Green Island, they said. That was all too hard for us. What do they think we are, bloody tourists or something! We'll just give Green Island a miss, thanks very much.

Time was passing. Port Douglas was going to be our next stop. There was a new Mirage luxury resort there, but due to the current pilots' strike, there were no planes flying, hence no guests. It was only a day sail from Cairns with a fresh trade wind, and we entered the Dickson Inlet in the late afternoon. The marina wasn't all that expensive, and we had a few very pleasant days' there.

Just off Port Douglas, 15kms out, are The Low Isles. They are well known by yachties and provide one of the best anchorages around. It was good to get away from the fleshpots, and have some peace and quiet, in a great anchorage.

Then on to Cooktown, immortalised by Captain Cook, who'd beached 'Endeavour' there for repairs. It was a trip of more than 100kms, and we didn't want to arrive there after dark, so we made it an overnight trip, and arrived in Cooktown in daylight. The course was straightforward enough, but we were now entering the northern part of the Great Barrier Reef, where the reef comes in much closer to the coast, and the channels are narrower. There's quite a lot of shipping using these channels. You don't notice it farther south, where there is more room, but it was brought sharply to our attention on this occasion when two ships came down, one after the other, from the north just as we were trying to negotiate the narrowest part of the channel. They have right of way, of course, and we had an interesting time keeping out of their way at just about the exact spot where the 'Endeavour' ran aground. I should mention here that fishing boats mostly operate at night, and also have right of way, especially when they are trawling.

We were fairly exhausted when we anchored in the

Chapter 7 - Consequence

Endeavour River next morning. We had anchored quite close to the wharf and getting ashore was easy. We spent a few days exploring the town, climbed Mount Cook (around 500m), and visited Cooktown's five pubs. There weren't any other cruising yachts around, so we had the anchorage more or less to ourselves.

The trade wind had picked up and was blowing strongly, and at this time of year it might continue for weeks. Our next destination was going to be Lizard Island but exiting the Endeavour River was going to be tough. It was no good waiting around for the weather to ease; at this time of year, that could be a week or more. We just had to bite the bullet and get on with it. I remembered, back in 1977, it was just the same; we came out of the Endeavour River with the sails ready to go up. I'd already put one reef in the main, and I had the yankee as a headsail. I got Chloe to take the tiller while I got the sails up. It was hairy stuff, worse than crossing Moreton Bay. The wind was more east than sou'east which meant we couldn't lay Cape Bedford, about 30kms north, and had to tack and tack and tack and tack. Finally, after about four hours, we rounded the cape, and were able to ease sheets, and race off towards Lizard Island.

As we approached Lizard Island. There are dangers, and the sailing directions warns that not all of them are marked. Sure enough, as we were sailing along, I saw a bombie out to starboard (a bombora, or bombie, is a large coral head just below the surface, and does not break, so it's often hard to spot). We'd just passed that one when I saw another one straight ahead. A wild swing to port and we missed it. Thankfully, there were no more, and it was a relief when we rounded the point and anchored in Watsons Bay. (By now, we were a long way outside the shipping channel and these hazards don't get the same attention that they would were they in the channel). Chloe had gone pretty quiet.

DESTINATION CAPE **HORN**

She'd done it pretty tough coming out of Cooktown. I tried to explain to her that sometimes cruising is like that; it's just a challenge that you have to meet. I started to talk to her about our next move, which would be to Thursday Island. We would sail in daylight, and anchor overnight. There were plenty of suitable anchorages. It would take about a week to get there. I had done the same voyage single-handed in 1977 with the help of Alan Lucas's book, 'Cruising the Coral Coast'.

'I don't want to go,' she said.

'I'm scared I might have a breakdown.'

I knew that she had a nervous disposition, and she had had a breakdown a few years earlier when her marriage broke down.

'I thought we were going round Australia,' I said.

'I'm sorry… I can't go on… But you can go on. I'll make my way back to Sydney.'

I couldn't abandon her. In the end, we decided that Chloe would fly to Cairns, and check herself into hospital, get herself sorted out, and I would bring 'Consequence' back to Cairns. Then, when she was ready, we would sail back to Sydney.

I was disappointed. To circumnavigate Australia would be no small achievement, and it would have been a great shakedown for

'Consequence', still a fairly new boat. Then I would have been ready for the Southern Ocean, and Cape Horn, the ultimate challenge.

We went to see the people at the resort, and yes, she could fly back to Cairns the following afternoon. So that was it; she flew out, and I set sail the next morning. There would be no stops, just straight through. There was a good favourable breeze to start with, and I made good progress in the first 24 hours, but the breeze eased right away, and I ended up having to start the

Chapter 7 - Consequence

auxiliary. It was just over 48 hours when I anchored in Trinity Inlet opposite the yacht club. A brief rest and a cup of coffee, then ashore.

I didn't know where the hospital was, but I soon found it, and there was the Dear Lady, all smiles, and feeling much better, she said.

'The doctors want me to stay a bit longer,' she said, 'Maybe tomorrow they'll let me go.'

'OK, that's fine,' I said, 'but let's just sit down, and make some plans.

'We're going back to Sydney, but let's stop at places where we didn't stop at on the way up.'

For example, we hadn't stopped at Bowen on the way up, 'so when we leave here, let's go straight to Bowen,' I suggested.

'We'll be at sea for two nights and arrive in Bowen on the third day; rest up for a few days, and leave Bowen and pass through the Whitsundays, and plan to stop at maybe Yepoon. Rest up for a few days more and try for Gladstone; a good rest in Gladstone, then try for Manly boat harbour in Brisbane.'

From Manly, we'd go down through the inland waterways to Southport.

Then Southport to Coffs Harbour to see the family, and finally Coffs Harbour to Sydney, and back to Balmain. We agreed that that would be the way to go and would get us back to Balmain sometime in

November. It had crossed my mind that if we could get back to Sydney by November, maybe I could be ready to take on the Southern Ocean before Christmas. I was a bit optimistic as it worked out, but I had forgotten what the main purpose of this new boat was.

Chloe was released from hospital the following day, and we

had a couple of quiet days getting ready, upping the anchor, and setting sail on our way south. The weather was good, with light to moderate breezes. Overnight there were minor problems with shipping, and fishing boats, but the voyage was mostly enjoyable, and we arrived in Bowen after lunch on the third day.

There is a boat harbour at Bowen, and we were lucky enough to get a

marina berth. Bowen is a large town, population around 10,000. It's the centre of a mango producing area (mango is our favourite fruit). The farmers were looking for workers, with the picking season just about to start, but it would have gone on until after Christmas, which was far too late for us.

We were in Bowen when the Rugby League Grand Final was played in

Sydney. We went to watch it at the yacht club. My team, the Balmain

Tigers, had made it through to the final, and were playing the Canberra Raiders, a new team to the competition, who had never won it before. The Raiders also featured a number of Queensland players, so the locals around Bowen were barracking for them. The game was exiting and close, and with just a few minutes to go, the scores were level. Then came an opportunity for the Tigers. They were attacking the Raiders line, and on the last tackle (Backdoor) Benny Elias took a pot shot at field goal. The ball hit the crossbar, and went straight up in the air, then came back down on the wrong side, meaning that there was no score. The scores were still tied at fulltime, and so we went into extra time. Sadly, the Tigers were spent; they had no more to give. The Raiders scored two tries and won the game. I think the Dear Lady and I were the only two Tigers supporters in the club, and we got a good razzing as we left. Just one of those things that I

Chapter 7 - Consequence

remember vividly to this day.

Gladstone was supposed to be our next destination, but it was a bit ambitious. Maybe we should have planned for Mackay, but we pressed on until, late in the afternoon, it was obvious that we weren't going to make it. We were close to Great Keppel Island, off Rockhampton, and that was a reasonable safe anchorage, a little bit of movement, but not too bad.

Next day, we went ashore for a walk. There was a resort on the island. Seemed like a fairly old place, not nearly as upmarket as some that we'd seen. We had a couple of drinks in the bar, then retreated back to 'Consequence'.

We got started bright and early next morning. The access to Gladstone was tricky, and we didn't want to get caught out again. As it was, it was well into the afternoon before we came around East Point on Facing Island and negotiated the narrow channel into Port Curtis. Gladstone is an industrial city; even the fishing industry has been pushed to one side. But it does have a very nice marina at reasonable cost, and we stayed there for a few days. While we were there, that one of my investments matured, and I was able to get to the bank, and renew it for a year at 19.5 percent. As I sit here typing this in 2021, when interest rates are close to zero, I can reflect on how easy it was to make money back then.

It was hot while we were in Gladstone, and we realised that the seasons were changing, and we had to get a move on, and get out of these northern waters before the summer set in.

We wanted to get next to Manly boat harbour in Moreton Bay. We were not going to go through the Great Sandy Strait this time, but down the seaward side of Fraser Island. That meant we had to swing well wide of the Breaksea Spit, which was well marked with a lightship, and also Sandy Cape shoal, which wasn't marked at all. It was just over 500kms to Manly, and that meant at least two nights at sea. The weather remained fine for us, and the dangerous area around Sandy Cape was negotiated safely. We crossed Moreton Bay, a dangerous area in its own right, and we arrived safely in Manly in the afternoon of the third day. We managed to get a berth in one of the marinas there and rested up for a few days. Manly is a suburb of Brisbane, and so has every facility one might need.

Our next destination, Southport, on the Gold Coast, was equally civilised. We'd decided that we were going to take the 'short cut' to Southport, through a marked channel that cuts through the mangrove swamps from the south of Moreton Bay to The Broadwater at Southport. It saves a long detour and two dangerous bars. We went through the channel, and an interesting trip it was. We got through to the Broadwater, then down to Southport late in the afternoon, and just in time to get a berth at the first marina that we came to. It was a bit expensive, but it was just for a few days. The Gold Coast, known mainly for Surfers Paradise, is a bit of a playground, and attracts people from all over the world with its climate, white beaches, and easy-going lifestyle. Like most places of this type, it's very expensive, and the Dear Lady and I had a day or two checking out some of the well-known places, and then were very happy to leave.

We were at that time of year when the daily sea breeze had set in, and so between the sea breeze, and the south setting current we were in Coffs Harbour in a couple of days, racing into the harbour ahead of a freshening nor'-easter. We found a marina berth, no problem.

Jimmy and Karen had turned up on Impossible', and we rented a car, and toured around the places of Chloe's youth. Met family and many others, and suddenly it was time to go. The weather forecast looked good, so we set off. The weather held,

Chapter 7 - Consequence

and three and a bit days later we sailed through Sydney Heads. I'd called the people at Balmain on the sea phone on the way down the coast, and they told us there would be a berth for us when we arrived. That was good news, and we made straight for Balmain.

This was now early December, and I'd known for a couple of weeks that

I wouldn't have time to get the boat ready and set off on the Southern Ocean voyage this year. This meant I had about 10 months to get myself and the boat ready to leave in early December 1990. The first thing to do was to settle down and go back to work.

Chloe had gone back to work, as well. She'd been welcomed back as a telephonist at Telecom Australia. So, we settled down in Sydney for the time being, still living on board the boat. This continued until about May, and she was talking about moving off the boat for the winter. One day, she put an idea to me.

'I've got some money left over from the divorce settlement,' she said,

'and I want to buy a house.'

I'd been half expecting this for some time. My position was that I had the Southern Ocean trip coming up, but after that, maybe a house would be a good idea. So, we talked around it, and decided that we wanted to stay together, and we would buy a house around Balmain not far from the marina, within walking distance. We found a terrace house that we both liked, and I put up half the money, and she put in what she had, and she had to borrow a relatively small amount. Interest rates were falling but were still historically high. But she was working and could cope with it.

When I wasn't working, I was on the boat, just picking over

everything very carefully. I'd decided to go for the GPS system of navigation. It had superseded the old Satnav system and was far better with all its extra satellites. They were quite reasonably priced, and I bought two, one of which would stay unopened, left in its sealed container in case of emergency. The other one, I set up in the navigation area, ready to use. I still had a sextant, and nautical almanac, again, just in case of emergency. I also invested in a first-class HF radio. Chloe was insisting that I report my position regularly, and the only way to do that is by HF. After it was installed and properly set up, I took the boat offshore and tested it by calling coastal stations up and down the coast, and in New Zealand. They all gave me a good report, which gave me a confidence boost.

I kept working until the October holidays. From then on, I was fully involved in preparations for the voyage. We had a get-together with the crowd from Taren Point. They all wished me well. I was starting to feel the pressure. I always felt pressure before a major voyage, so that wasn't new. I can remember right back to my schooldays, feeling pressure before final exams. The more pressure I felt, the better I did in the exam. Was that an omen?

The day came; it was time to go. The weather report was good, for the next few days, anyway. After that, it's in the lap of the gods. We leave Balmain and head down to Neutral Bay, customs formalities, passport stamped, then down the harbour, through the heads, turn south. The plan was to follow the great circle route, get down south, right into the Roaring 40s, south of NZ, and stay in the 40s from there on. The weather was light for the first couple of days, but we did get a bit of help from the current.

We were somewhere off Montague Island when the weather

Chapter 7 - Consequence

changed.

The barometer had been falling, and it looked like a thunderstorm was building. The storm came overnight, and I had to drop the sails for a while till the gusty winds settled down. By daylight, the storm had disappeared, and the wind had shifted to the west then sou'-west at 30 knots, and we were bashing into it, under much reduced sail. Now we were out into Bass Strait, that notorious stretch of water, and it seemed that the weather had settled for the time being, and this is what we could expect for some days, most probably. Eventually the wind would shift to the south, then the sou'-east, but for now we had to cope with sou'-west.

As the day wore on, the wind shifted sou'-sou'-west and increased to 35 knots-plus. Conditions were horrendous; the seas were short, and every swell had a top on it. We were just going bash, bash, bash into the short seas. Spray was flying everywhere, and the wind was shrieking in the rigging. Cooking an evening meal was out of the question; just a cup of coffee and a hard tack bikkie.

It was that time of day when I'd get the rum bottle out and have a couple of slugs. But it didn't help. It made it worse. I started to have all sorts of negative thoughts. I tried to lie down and sleep, perhaps I'd get a bit of rest, if not sleep. Why am I doing this? Because I stuffed it up the first time. Memories come back, not good. I suffered nightmares for a

few weeks after Cape Horn, then they died out, and never bothered me again, till now. Get a grip man; get a grip. Hang in there. Morning comes, everything is grey. We're still bashing along. We're well off the Tasmanian coast, not far enough south to be feeling the Southern Ocean. If we could hold this course, it would take us to Stewart Island, at the bottom of NZ. We need to round it, of course. About 90kms sou'-west of Stewart Island is Broughton Island. Just about halfway between the two would be great. Just get through the day, this wind must ease sooner or later, or shift direction.

I get through the day, but the night I can't handle… rum, and more rum. I'm an absolute mess. I can't go on. Turn back, my mind has been telling me for hours now. It's done. In the wee small hours of the morning. I just turned her around, reset the sails, reset the self-steering gear, and we're headed back to Sydney. I'm shot to bits. How can I justify turning back?

I called Chloe on the sea phone, on the way back up the NSW coast. She was non-committal shocked even.

She told me later that she was taken aback and didn't know what to say.

I called the Balmain Marina. They were a bit more sympathetic. Sorry it didn't work out for you, they said. Your berth is still here, and you can have it back. See you when you get here. The customs were quite sympathetic. Don't feel too bad about it, they said. We get people from time to time, who discover too late that the sea is not for them.

I went back to work. That was the best thing I could have done. The public don't know anything about my problems, and they don't want to know. Working takes me away from all the bad memories. The mind focuses on the present, and the bad thoughts fade into the background.

I didn't sell the boat straight away, there were other plans, ideas, attempts. It took me some time to understand that the world had moved on, and I was never going to do any serious sailing again. The boat was just sitting in its marina berth at $200 a month, never mind all the other basic maintenance, and that couldn't continue. So, I sold her, for about half of the cost to build

Chapter 7 - Consequence

and equip. It was never about money, of course. It was never an investment. It was about sailing, and I stuffed that up good and proper. I was lucky to get what I got for her in the end.

In 1996, we bought a new house in Hunters Hill. Chloe's daughter, Shelley, came in with us, and that made it more affordable for all.

I kept working, which I enjoyed as much as anything else, and having a bit of cash income was great. I wanted to continue working until I was 80, but life intervened again. The summer of 2019-20 brought disastrous bushfires, and the Sydney basin was surrounded. The smoke from these fires collected in the basin. Some days, it was so thick we could hardly see the sun. The smoke was poisonous, and people were being hospitalised. I kept working, but it was pretty tough going. At long last, we had widespread heavy rain, which put out the fires. The end of the bushfires just about coincided with the arrival of the first Covid-19 cases, and it wasn't long before cases were increasing rapidly. A cruise ship arrived in Sydney Harbour (the last before the industry closed down), and disgorged hundreds of passengers, all carrying the virus, and spreading it far and wide.

This was the final straw for me. Business had declined to the point where I was earning next to nothing, and there was a rapidly increasing chance that I would get the virus myself and pass it on to Chlöe and Shelley. I stopped work, the only sensible thing to do. I'll give it six months, I thought, and see how it looks. In that time, the situation deteriorated into a world-wide pandemic, and I couldn't see any improvement any time soon.

One day, the base called me, and a nice lady asked me when I was coming back to work.

'We're expecting things to improve shortly,' she said. 'We need people like you.'

'Oh really,' I said, and hummed and hawed and maybe'd.

'Look,' she said. 'If you are not going to come back, we're going to say that you've retired, and we're going to cancel your registration, etc.' 'Ok,' I said. 'I hear what you say. It looks like I'm retired then.'

I miss working. I miss the cut and thrust, and above all I miss the people, bless their souls. I could have gone on for long enough, but the pandemic ruled otherwise.

People have been saying to me for years that I should write about my sailing exploits. It wasn't something that I was all that keen on, and anyway I never seemed to have the time. Countless thousands have circumnavigated, and many of them have written about it. I doubt if there is any aspect of it that hasn't been well-covered. I saw it as a personal challenge to sail and navigate safely around the world single handed via the Panama Canal. I visited most of the usual places, meeting new people every place that I went, and having a great time doing it. When I left Sydney in 1981, it was to sail around the world in the other direction, and via the Great Capes. That was the challenge. As we approached Cape Horn, we experienced by far the worst weather and sea conditions we'd ever been through. We survived that, and looked set to round the cape, but I relaxed and allowed myself to fall asleep. That was inexcusable, and I had no right to survive the shipwreck that followed. The escape was amazing, but I didn't deserve it. These events I will live with for the rest of my life.

Now that I'm retired, and have time on my hands, I've turned to writing, not because I seek notoriety, or money, but because it gives me something to do, and for my family, and one or two others, so they may not judge me too harshly. Chloe and I, and Shelley, live a relatively comfortable life. We're not millionaires, but we have a nice house (which we own), in a nice

Chapter 7 - Consequence

suburb, with nice neighbours, and an income to keep us going for many years yet.

Milton Keynes UK
Ingram Content Group UK Ltd.
UKHW020740151223
434437UK00010B/531